# 2D Materials for Infrared and Terahertz Detectors

# Series in Materials Science and Engineering

The series publishes cutting edge monographs and foundational textbooks for interdisciplinary materials science and engineering. It is aimed at undergraduate and graduate level students, as well as practicing scientists and engineers. Its purpose is to address the connections between properties, structure, synthesis, processing, characterization, and performance of materials.

Skyrmions: Topological Structures, Properties, and Applications
*J. Ping Liu, Zhidong Zhang, Guoping Zhao, Eds.*

Computational Modeling of Inorganic Nanomaterials
*Stefan T. Bromley, Martijn A. Zwijnenburg, Eds.*

Physical Methods for Materials Characterisation, Third Edition
*Peter E. J. Flewitt, Robert K. Wild*

Conductive Polymers: Electrical Interactions in Cell Biology and Medicine
*Ze Zhang, Mahmoud Rouabhia, Simon E. Moulton, Eds.*

Silicon Nanomaterials Sourcebook, Two-Volume Set
*Klaus D. Sattler, Ed.*

Advanced Thermoelectrics: Materials, Contacts, Devices, and Systems
*Zhifeng Ren, Yucheng Lan, Qinyong Zhang*

Fundamentals of Ceramics, Second Edition
*Michel Barsoum*

Flame Retardant Polymeric Materials, A Handbook
*Xin Wang and Yuan Hu*

2D Materials for Infrared and Terahertz Detectors
*Antoni Rogalski*

# Series Preface

The series publishes cutting edge monographs and foundational textbooks for interdisciplinary materials science and engineering.

Its purpose is to address the connections between properties, structure, synthesis, processing, characterization, and performance of materials. The subject matter of individual volumes spans fundamental theory, computational modeling, and experimental methods used for design, modeling, and practical applications. The series encompasses thin films, surfaces, and interfaces, and the full spectrum of material types, including biomaterials, energy materials, metals, semiconductors, optoelectronic materials, ceramics, magnetic materials, superconductors, nanomaterials, composites, and polymers.

It is aimed at undergraduate and graduate level students, as well as practicing scientists and engineers.

*Proposals for new volumes in the series may be directed to Carolina Antunes, Commissioning Editor at CRC Press, Taylor & Francis Group (Carolina.Antunes@tandf.co.uk).*

# 2D Materials for Infrared and Terahertz Detectors

Antoni Rogalski

**CRC Press**
Taylor & Francis Group
Boca Raton  London  New York

CRC Press is an imprint of the
Taylor & Francis Group, an **informa** business

First edition published 2021
by CRC Press
6000 Broken Sound Parkway NW, Suite 300, Boca Raton, FL 33487-2742

and by CRC Press
2 Park Square, Milton Park, Abingdon, Oxon, OX14 4RN

*Library of Congress Cataloging-in-Publication Data*

Names: Rogalski, Antoni, author.
Title: 2D materials for infrared and terahertz detectors / Antoni Rogalski.
Other titles: Two-dimensional materials for infrared and terahertz detectors
Description: First edition. | Boca Raton, FL : CRC Press, Taylor & Francis Group, 2020. | Series: Series in materials science and engineering | Includes bibliographical references and index.
Identifiers: LCCN 2020020959 (print) | LCCN 2020020960 (ebook) | ISBN 9780367477417 (hardback) | ISBN 9781003043751 (ebook)
Subjects: LCSH: Infrared detectors--Materials. | Microwave detectors--Materials. | Thin film devices--Materials. | submillimeter waves. | Graphite.
Classification: LCC TA1573 .R638 2020 (print) | LCC TA1573 (ebook) | DDC 621.36/20284--dc23
LC record available at https://lccn.loc.gov/2020020959
LC ebook record available at https://lccn.loc.gov/2020020960

ISBN: 9780367477417 (hbk)
ISBN: 9781003043751 (ebk)

Typeset in Minion
by Deanta Global Publishing Services, Chennai, India

# Contents

# Author

**Antoni Rogalski** is a professor at the Institute of Applied Physics, Military University of Technology in Warsaw, Poland. He is one of the world's leading researchers in the field of infrared (IR) optoelectronics. He has made pioneering contributions in the areas of theory, design, and technology of different types of IR detectors. In 1997, Professor Rogalski received an award from the Foundation for Polish Science, the most prestigious scientific award in Poland, for his achievements in the study of ternary alloy systems for infrared detectors. His monumental monograph, *Infrared and Terahertz Detectors* (published in three editions by Taylor and Francis), has been translated into Russian and Chinese. In 2013, Professor Rogalski was elected as an Ordinary Member of the Polish Academy of Sciences and as a member of the Central Commission for Academic Degrees and Titles. Since early 2015, he has been the Dean of the Faculty of Technical Sciences of the Polish Academy of Sciences, and, from 2016, he has been a member of the group for affairs of scientific awards of the Prime Minister of Poland.

Professor Rogalski is a fellow of the International Society for Optical Engineering (SPIE), vice-president of the Polish Optoelectronic Committee, editor-in-chief of the journal *Opto-Electronics Review* (1997-2015), deputy editor-in-chief of the *Bulletin of the Polish Academy of Sciences: Technical Sciences* (2003-present), and a member of the editorial boards of several international journals. He is an active member of the international technical community – as chair and co-chair, organizer, and member of scientific committees of many national and international conferences on optoelectronic devices and materials sciences.

# Preface

SINCE THE DISCOVERY OF graphene, its applications to electronic and optoelectronic devices have been intensively and thoroughly researched. Its extraordinary and unusual electronic and optical properties allow graphene and other two-dimensional (2D) materials to be promising candidates for infrared (IR) and terahertz (THz) photodetectors. Until now, however, their place in the wide-infrared detector family has not been evaluated and this topic is generally omitted from the review literature.

The main goal of this book is to provide a critical view on the present state of 2D-material photodetector technologies and on future developments with respect to global competition with existing industrially mature material detector systems, such as HgCdTe, InGaAs, type-II superlattice III-V compounds, and microbolometers. This book also considers the challenges facing development of focal plane arrays for the future. Special attention is paid toward the main trends in development of arrays in the near future, such as increases in pixel count to above $10^8$ pixels, with pixel size decreasing to about 5-µm, mostly for uncooled infrared arrays. Until now, these questions have not been considered in literature reviews devoted to 2D-material IR and THz detectors.

Most of the 2D layered semiconducting material photodetectors operate at the visible and near-infrared regions. However, the thrust of this book is mainly directed to effective IR and THz detectors, based on 2D materials. This book

- gives brief accounts of the different types of 2D materials used in the fabrication of IR and THz detectors,

- describes advantages and disadvantages of 2D materials as IR and THz detectors,

- sets in order the performance of 2D material IR and THz detectors among the family of common commercially available detectors,

- tries to predict the future role for 2D materials in the family of detectors,

- predicts the main trends in development of arrays in the near future.

The number of published papers devoted to the use of 2D materials as sensors is huge. However, the authors of these papers mainly address their work to researchers involved in investigations of 2D materials. In this book, the position of 2D material detectors is considered in comparison with the present state of conventional infrared and terahertz detectors offered on the global market. In this way, the book gives an overview of the performance of emerging 2D material detectors, comparing them with traditionally and commercially available ones under different conditions, including high operating-temperature conditions.

This monograph is divided into eight chapters. After the introduction, two chapters (2 and 3) describe detector characterization and fundamentals of detection mechanisms for both thermal and photon detectors, including detector performance limits. These initial chapters provide a tutorial introduction to the technical topics that are necessary for a thorough understanding of the different types of detectors and systems. In Chapter 3, a new reference benchmark, the so-called "Rule 19", is introduced for prediction of the performance of background-limited HgCdTe photodiodes, operated near room temperature. This rule is subsequently addressed in the following chapters (6, 7, and 8) as a benchmark against which to compare alternative 2D material technologies.

In Chapter 4, topics are considered which are almost completely omitted by the scientific community researching 2D detector materials, including future trends in the development of focal plane arrays. Taking into account the early stages of development and manufacturability, such considerations are essential to make a realistic assessment of the prospects for subsequent commercialization of 2D-material photodetectors.

The next four chapters (5, 6, 7, and 8) briefly describe the fundamental properties of graphene-based materials and other 2D materials, and the performance parameters (such as responsivity, detectivity, and response time) of detectors fabricated with these materials, in comparisons between 2D material-based detectors and traditional detectors on the global market, including both experimental data and theoretical predictions. Final conclusions predict the likely place of 2D material-based detectors in the wide-IR detector family, in the near future.

The presentation level of this book is suitable for graduate students in physics and engineering, who have received background training in modern solid-state physics and electronic circuits. This book would also be of interest to individuals working with aerospace sensors and systems, remote sensing, thermal imaging, military imaging, optical telecommunications, infrared spectroscopy, and light detection and ranging

This book, I hope, will provide a timely and appropriate analysis of the latest developments in 2D- material infrared and THz detector technology and a basic insight into the fundamental processes important to evolving detection techniques. The book covers different types of detectors, including the relevant aspects of theory, types of materials, their physical properties, and detector fabrication.

**Antoni Rogalski**

# Acknowledgments

IN THE COURSE OF writing this book, many people have assisted me and offered their support. I would like to express appreciation to the management of the Institute of Applied Physics, Military University of Technology, for providing the environment in which I worked on the book. The writing of the book has been partially done under financial support of the The National Science Centre (Poland) – (Grant nos. UMO-2018/30/M/ST7/00174; UMO-2018/31/B/ST7/01541; and UMO-2019/33/B/ST7/00614).

# CHAPTER 1

# Introduction

INFRARED (IR) RADIATION ITSELF was unknown until 220 years ago, when Herschel's experiment with the thermometer was first reported. The first detector consisted of a liquid in a glass thermometer with a specially blackened bulb, to absorb radiation. Herschel built a crude monochromator that used a thermometer as a detector, so that he could measure the distribution of energy in sunlight [1].

The early history of IR was reviewed about 60 years ago in two well-known monographs [2,3]. Much historical information can be also found in more recently published papers [4,5]. The initial infrared detectors were based on the class of thermal detectors: thermometers, thermocouples, and bolometers [6]. In 1821, T.J. Seebeck discovered the thermoelectric effect, and soon afterward, in 1829, L. Nobili created the first thermocouple. In 1833, M. Melloni modified the thermocouple and used bismuth and antimony for its design [7]. Then, in 1835, Nobili, together with Melloni, constructed a thermopile capable of sensing a person 10 m away. The third type of thermal detector, the bolometer/thermistor, was invented by S.P. Langley in 1878. By 1900, his bolometer was 400 times more sensitive than his first efforts, and his latest bolometer could detect the heat from a cow at a distance of ¼ mile [8].

The photoconductive effect was discovered by W. Smith in 1873, when he experimented with selenium as an insulator for submarine cables [9]. This discovery provided a fertile field of investigation for several decades, though most of the effort was of doubtful quality. By 1927, over 1500 articles and 100 patents had been published on photosensitive selenium [10]. Work on the IR photovoltaic effect in naturally occurring lead sulfide, or galena, was first published by Bose in 1904 [11]; however, the IR photovoltaic effect was not exploited in a radiation detector for several more decades.

The photon detectors were developed in the twentieth century. The first IR photoconductor was developed by T.W. Case in 1917 [12]. He discovered that a substance composed of thallium and sulfur exhibited photoconductivity. Later, he found that the addition of oxygen greatly enhanced the response [13]. However, the instability of the resistance in the presence of light or polarizing voltage, the loss of responsivity due to overexposure to light or high noise, its sluggish response, and the lack of reproducibility seemed to be inherent weaknesses.

Since about 1930, the development of IR technology has been dominated by photon detectors. In about 1930, the appearance of the Cs-O-Ag phototube, with more stable characteristics, discouraged further development of photoconductive cells to a great extent until about 1940. At that time, interest in improved detectors had begun [14,15]. In 1933, Kutzscher, at the University of Berlin, discovered that lead sulfide (from natural galena found in Sardinia) was photoconductive and had a response to about 3 μm. This work was, of course, carried out under great secrecy and the results were not generally known until after 1945. Lead sulfide was the first practical IR detector deployed in a variety of applications during the war. In 1941, Cashman improved the technology of thallous sulfide detectors, which led to successful production [16]. After success with thallous sulfide detectors, Cashman concentrated his efforts on lead sulfide and, after World War II, found that other semiconductors of the lead salt family (PbSe and PbTe) showed promise as IR detectors [17]. Lead sulfide photoconductors were brought to the manufacturing stage of development in Germany in about 1943. They were first produced in the United States at Northwestern University, Evanston, Illinois in 1944 and, in 1945, at the Admiralty Research Laboratory in England [17].

Many materials have been investigated in the IR field. Observing a history of the development of the IR detector technology, a simple theorem, after Norton [18], can be stated: "All physical phenomena in the range of about 0.1–1 eV can be proposed for IR detectors." Among these effects are: thermoelectric power (thermocouples), change in electrical conductivity (bolometers), gas expansion (the Golay cell), pyroelectricity (pyroelectric detectors), photon drag, the Josephson effect (Josephson junctions, SQUIDs), internal emission (PtSi Schottky barriers), fundamental absorption (intrinsic photodetectors), impurity absorption (extrinsic photodetectors), low-dimensional solids [superlattice (SL), quantum well (QW), and quantum dot (QD) detectors], different types of phase transitions, and so on.

Figure 1.1 gives approximate dates for significant developments for the materials mentioned. The years during World War II saw the origins of modern IR detector technology, supported by the discovery of the transistor in 1947 by W. Shockley, J. Bardeen, and W. Brattain [19]. Recent success in applying IR technology to remote sensing problems has been made possible by the successful development of high-performance IR detectors over the past seven decades. Photon IR technology, combined with semiconductor material science, photolithography technology developed for integrated circuits, and the impetus of Cold War military preparedness, propelled extraordinary advances in IR capabilities within a short period of time during the past century [20].

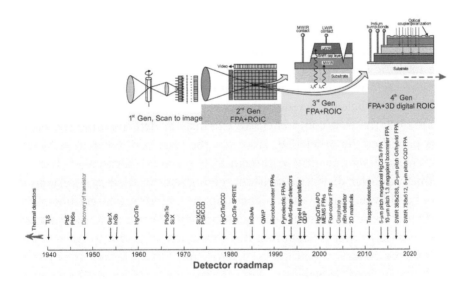

FIGURE 1.1 The history of the development of IR detectors and systems. For principal military and civilian applications, four generation systems can be considered: first-generation (scanning systems), second-generation (staring systems, electronically scanned), third-generation (staring systems, with large number of pixels and two-color functionality), and fourth-generation (staring systems with very large number of pixels, multi-color functionality, 3D ROIC, and other on-chip functions) systems, offering other functions, e.g. better radiation/pixel coupling, avalanche multiplication in pixels, and polarization/phase sensitivity.

## 1.1 HISTORICAL ASPECTS OF MODERN INFRARED TECHNOLOGY

During the 1950s, IR detectors were built using single-element-cooled lead salt detectors, primarily for anti-air missile seekers. Usually, lead salt detectors were polycrystalline and were produced by vacuum evaporation and chemical deposition from a solution, followed by a post-growth sensitization process [17]. The preparation process of lead salt photoconductive detectors was usually not well understood, and reproducibility could be achieved only after following well-tried recipes. The first extrinsic photoconductive detectors were reported in the early 1950s [21], after the discovery of the transistor, which stimulated a considerable improvement in the growth and material purification techniques. Since the techniques for controlled introduction of impurities became available for germanium earlier, the first high-performance extrinsic detectors were based on germanium. Extrinsic, photoconductive response from copper, zinc, and gold impurity levels in germanium gave rise to devices using the 8- to 14-μm longwave IR (LWIR) spectral window and beyond to the 14- to 30-μm very longwave IR (VLWIR) region. The extrinsic photoconductors were widely used at wavelengths beyond 10 μm, prior to the development of the intrinsic detectors. They must be operated at lower temperatures to achieve performance similar to that of intrinsic detectors, and a sacrifice in quantum efficiency is required to avoid thick detectors.

In 1967, the first comprehensive extrinsic Si detector-oriented paper was published, by Soref [22]. However, the state of the extrinsic Si was not changed significantly. Although Si has several advantages over Ge (namely, a lower dielectric constant, giving shorter dielectric relaxation times and lower capacitance, higher dopant solubility, a larger photoionization cross section for greater quantum efficiency, and a lower refractive index for lower reflectance), these were not sufficient to warrant the necessary development efforts needed to bring it to the level of the by-then highly developed Ge detectors. After the concept lay dormant for about 10 years, extrinsic Si was reconsidered after the invention of charge-coupled devices (CCDs) by Boyle and Smith [23]. In 1973, Shepherd and Yang [24] proposed the metal-silicide/silicon Schottky barrier detectors. For the first time, it became possible to have much more sophisticated readout schemes, so that both detection and readout could be implemented in one common silicon chip.

At the same time, rapid advances were being made in narrow-bandgap semiconductors, that would later prove useful in extending wavelength capabilities and improving sensitivity. The first such material was InSb, a member of the newly discovered III-V compound semiconductor family. The interest in InSb stemmed, not only from its small energy gap, but also from the fact that it could be prepared in single crystal form, using a conventional technique. The end of the 1950s and the beginning of the 1960s saw the introduction of narrow-gap semiconductor alloys in III–V ($InAs_{1-x}Sb_x$), IV–VI ($Pb_{1-x}Sn_xTe$), and II–VI ($Hg_{1-x}Cd_xTe$) material systems. These alloys allowed the bandgap of the semiconductor, and hence the spectral response of the detector, to be custom tailored for specific applications. In 1959, research by Lawson and coworkers [25] triggered the development of variable-bandgap $Hg_{1-x}Cd_xTe$ (HgCdTe) alloys, providing an unprecedented degree of freedom in IR detector design. This first paper [25] reported both photoconductive and photo-voltaic response, extending out to 12 μm in wavelength. Soon thereafter, working under a U.S. Air Force contract with the objective of devising an 8–12 μm background-limited semiconductor IR detector that would operate at temperatures as high as 77 K, the group, led by Kruse, at the Honeywell Corporate Research Center in Hopkins, MN, developed a modified Bridgman crystal growth technique for HgCdTe. They soon reported both photoconductive and photovoltaic detection in rudimentary HgCdTe devices [26].

The fundamental properties of narrow-bandgap semiconductors (high optical absorption coefficient, high electron mobility, and low thermal generation rate), together with the capability for bandgap engineering, made these alloy systems almost ideal for a wide range of IR detectors. The difficulties in growing HgCdTe material, due significantly to the high vapor pressure of Hg, encouraged the development of alternative detector technologies over the past 40 years. One of these was PbSnTe, which was vigorously pursued in parallel with HgCdTe in the late 1960s and early 1970s [27–29]. PbSnTe was comparatively easy to grow, and high-quality LWIR photodiodes were readily demonstrated. However, in the late 1970s, two factors led to the abandonment of PbSnTe detector work: high dielectric constants and large mismatch of coefficient of thermal expansion (CTE) with Si. Scanned IR imaging systems of the 1970s required relatively fast response times, to avoid smearing the scanned image in the scan direction. With the trend today toward staring arrays, this consideration might be less important than it was when first-generation systems were being designed. The second drawback, a large CTE, can lead to failure of the indium bonds in hybrid structures (between the silicon readout and the detector array) after repeated thermal cycling, from room temperature to the cryogenic temperature of operation.

The material technology development was and continues to be primarily for military applications. In 1956, Texas Instruments had begun research on IR technology, which led to the signing of several contracts for a linear scanner, and subsequently to the invention of the first forward-looking infrared (FLIR) camera in 1963. As photolithography became available in the early 1960s, it was used to make IR detector arrays. Linear array technology was first applied to PbS, PbSe, and InSb detectors. The discovery in the early 1960s of extrinsic Hg-doped germanium [30] led to the first FLIR systems operating in the LWIR spectral window, using linear arrays. Because the detection mechanism was based on an extrinsic excitation, it required a two-stage cooler to operate at 25 K. The cooling requirements of intrinsic narrow- bandgap semiconductor detectors are much less stringent. Typically, to obtain the background-limited performance (BLIP), detectors for the 3–5 μm spectral region are operated at 200 K or less, while those for the 8–14 μm region are operated at the temperature of liquid nitrogen. In the late 1960s and early 1970s, the first-generation linear arrays of intrinsic HgCdTe photoconductive detectors were developed, in which an electrical contact for each element of a multielement array is brought off the cryogenically cooled focal plane to the outside, where there is one electronic channel at ambient temperature for each detector element Fig. 1.1). In 1972, Texas Instruments invented the HgCdTe Common Module concept, which contributed to a significant cost reduction and allowed for the reuse of common components. These allowed LWIR FLIR systems to operate with a single-stage cryoengine, making the systems much more compact and lighter, and consuming significantly less power.

Early assessment of the concept of the second-generation system showed that PtSi Schottky barriers, InSb, and HgCdTe photodiodes or high-impedance photoconductors, such as PbSe and PbS, and extrinsic silicon detectors were promising candidates because they had impedances suitable for interfacing with the field-effect transistor (FET) input of read-out multiplexes. Photoconductive HgCdTe detectors were not suitable due to their low impedance and high-power dissipation on the focal plane. A novel British invention, the SPRITE detector [31,32], extended conventional photoconductive HgCdTe detector technology by incorporating signal time delay and integration (TDI) within a single elongated detector element. Such a detector replaces a whole row of discrete elements of a conventional serial-scanned detector, external associated amplifiers, and time-delay circuitry. Although only used in small arrays of about 10 elements, these devices have been produced in the thousands.

In the late 1970s and through the 1980s, HgCdTe technology efforts focused almost exclusively on photovoltaic device development, because of the need for low power dissipation and high impedance in large arrays to interface with readout input circuits. The emergence of advanced epitaxial techniques [molecular beam epitaxy (MBE) and metalorganic chemical vapor deposition (MOCVD)], combined with the photolithography process, revolutionized the IR detector system industry, enabling the design and fabrication of complex focal plane arrays (FPAs). These efforts are finally paying off, with the birth of HgCdTe second-generation IR systems, that provide large two-dimensional (2D) arrays in both linear formats, with time delay and integration (TDI) for scanning imagers, and in square and rectangular formats for staring arrays. At the present stage of development, staring arrays have about $10^8$ elements and are scanned electronically by circuits integrated with the arrays. It is predicted that larger focal planes will be possible, constrained by budgets rather than by technology [33]. These 2D arrays of photodiodes, connected with indium bumps to a readout integrated circuit (ROIC) chip as a hybrid structure, are often called a sensor chip assembly (SCA).

The first megapixel hybrid HgCdTe FPAs were fabricated in the mid-1990s. However, present HgCdTe FPAs are limited by the yield of arrays, which increases their cost. In such a situation, alternative alloy systems for infrared detectors, such as quantum well infrared photodetectors (QWIPs) and type-II superlattices (T2SLs), are being evaluated.

Recently, considerable progress has been made toward III-V antimonide-based low-dimensional solid development and device design innovations. Their development results from two primary motivations: the perceived challenges of reproducibly fabricating high-operability HgCdTe FPAs at reasonable cost, and theoretical predictions of lower Auger recombination for T2SL detectors, compared with HgCdTe. Lower Auger recombination translates into a fundamental advantage for T2SL over HgCdTe in terms of lower dark current and/or higher operating temperatures, provided that other parameters, such as Shockley-Read-Hall lifetimes, are equal. Recently, Raytheon's III-V T2SL (type II superlattice)/nBn detectors have also reached a level of maturity that enabled the company to win the contract for the next-generation Distributed Aperture System (DAS) for the F-35 Joint Strike Fighter [34].

HgCdTe has inspired the development of the three "generations" of detector devices. Third-generation devices are defined here to encompass the more exotic device structures, embodied in two-color detectors and hyperspectral arrays, which are now in production programs. For example, Raytheon fabricates the eLRAS3 system (Long Range Scout Surveillance System), which provides the real-time ability to detect, recognize, identify, and geo-locate distant targets outside the zone of threat acquisition. This high-definition high-resolution FLIR (also called 3rd Gen FLIR) combines HgCdTe longwave and mid-wave infrared arrays.

The first three generations of imaging device systems rely primarily on planar FPAs. We are currently dealing with the fourth-generation staring systems, in which the main features are to be high resolution (with a very large number of pixels, above $10^8$), multi-color functionality, three-dimensional readout integration circuits (3D ROIC), and other integrated functions, e.g., better radiation/pixel coupling, avalanche multiplication in pixels, and polarization/phase sensitivity. The evolution of the fourth generation is inspired by the most famous visual systems, which are biological eyes. A solution, based on the Petzval-matched curvature, allowing the reduction of field curvature aberration, e.g. bonding the detectors to flexible or curved molds, has been proposed [35]. In addition, such a system combines such advantages as a simplified lens system, electronic eye systems, and wide field-of-view [36,37]. The colloidal quantum dot (CQD) [38] and 2D layered material photodetectors [39], fabricated on flexible substrates, are promising materials with which to overcome technical challenges in the development of fourth-generation IR systems. The unique and distinctive optoelectronic properties of graphene and related two-dimensional (2D) materials create a new platform for a variety of photonic applications, including infrared and terahertz photodetectors. In particular, there is growing interest in 2D materials for sensors, that have the potential to operate at room temperature.

As was mentioned previously, the development of IR technology has been dominated by photon detectors since about 1930. However, photon detectors require cryogenic cooling. This is necessary to prevent the heat generation by the charge carriers. The thermal transitions compete with the optical ones, making uncooled devices very noisy. The cooled thermal camera usually uses the Stirling cycle cooler, which is the most expensive component of the photon detector IR camera. Cooling requirements are the main obstacle to the widespread use of IR systems based on semiconductor photon detectors, making them bulky, heavy, expensive, and inconvenient to use.

The use of thermal detectors for IR imaging has been the subject of research and development for many decades. However, in comparison with photon detectors, thermal detectors have been considerably less exploited in commercial and military systems. The reason for this disparity is that thermal detectors are popularly believed to be rather slow and less sensitive in comparison with photon detectors. As a result, the worldwide effort to develop thermal detectors has been extremely small, relative to that of the photon detectors.

It must not be inferred from the preceding outline that work on thermal detectors has not been actively pursued. Indeed, some interesting and important developments have taken place along this line. In 1947, for example, Golay constructed an improved pneumatic infrared detector [40]. This gas thermometer has been used in spectrometers. The thermistor bolometer, originally developed by Bell Telephone Laboratories, has found widespread use in detecting radiation from low-temperature sources [41,42]. The superconducting effect has been used to make extremely sensitive bolometers.

Thermal detectors have also been used for infrared imaging. Evaporographs and absorption-edge image converters were among the first non-scanned IR imagers. Originally, an evaporograph was employed in which the radiation was focused onto a blackened membrane coated with a thin film of oil [43]. The differential rate of evaporation of the oil was proportional to the radiation intensity. The film was then illuminated with visible light to produce an interference pattern corresponding to the thermal picture. The second thermal-imaging device was the absorption-edge image converter [44]. Operation of this device was based upon utilizing the temperature dependence of the absorption edge of the semiconductor. The performance of both imaging devices was poor because of the very long time constraint and the poor spatial resolution. Despite numerous research initiatives and the attractions of ambient-temperature operation and low cost-potential, thermal detector technology has enjoyed limited success, in competition with cooled photon detectors, with respect to thermal imaging applications. A notable exception was the pyroelectric vidicon (PEV) [45], which was widely used by firefighting and emergency service organizations. The PEV tube can be considered analogous to the visible television camera tube, except that the photoconductive target is replaced by a pyroelectric detector and germanium faceplate. Compact, rugged PEV imagers have been offered for military applications but suffer the disadvantage of short tube-life and fragility, particularly the reticulated vidicon tubes, which are required for enhanced spatial resolution. The advent of the staring FPAs, however, marked the development of devices that would someday make uncooled systems practical for many, especially commercial, applications. The defining effort in this field was undertaken by Texas Instruments with contractual support from the Army Night Vision Laboratory [5]. The goal of this program was to build a staring FPA system based on ferroelectric detectors of barium strontium titanate. Throughout the 1980s and early 1990s, many other companies developed spatial devices based on various thermal detection principles.

The second revolution in thermal imaging began at the end of the twentieth century. The development of uncooled IR arrays, capable of imaging scenes at room temperature, has been an outstanding technical achievement. Much of the technology was developed under classified military contracts in the United States, so the public release of this information in 1992 surprised many in the worldwide IR community [46]. There has been an implicit assumption that only cryogenic photon detectors, operating in the 8–12 μm atmospheric window, had the necessary sensitivity to image objects at room temperature. Although thermal detectors have been little used in scanned imagers, because of their slow response, they are currently of considerable interest for 2D electronically addressed arrays, where the bandwidth is low and where the ability of thermal devices to integrate over a frame time is an advantage [47– 52]. Much recent research has focused on both hybrid and monolithic uncooled arrays and has yielded significant improvements in the detectivity of both bolometric and pyroelectric detector arrays. Honeywell has licensed bolometer technology to several companies for the development and production of uncooled FPAs for commercial and military systems. At present, compact megapixel microbolometer cameras are produced by Raytheon, L-3 Communications, FLIR, and DRS in the United States. The U.S. government allows these manufacturers to sell their devices to foreign countries, but not to divulge manufacturing technologies. Later on, several countries, including the United Kingdom, France, Japan, Israel, Korea, and China have "picked up the ball", determined to develop their own uncooled imaging systems. As a result, although the United States has a significant lead, some of the most exciting and promising developments for low-cost uncooled IR systems in the future may come from non-U.S. companies (e.g., microbolometer FPAs with series p-n junctions, developed by Mitsubishi Electric).

# REFERENCES

1. W. Herschel, "Experiments on the refrangibility of the invisible rays of the sun", *Philosophical Transactions of the Royal Society of London* **90**, 284 (1800).
2. R.A. Smith, F.E. Jones, and R.P. Chasmar, *The Detection and Measurement of Infrared Radiation*, Clarendon, Oxford, 1958.
3. P.W. Kruse, L.D. McGlauchlin, and R.B. McQuistan, *Elements of Infrared Technology*, Wiley, New York, 1962.
4. C. Corsi, "History highlights and future trends of infrared sensors", *Journal of Modern Optics* **57**, 1663–1686 (2009).
5. A. Rogalski, "History of infrared detectors", *Opto-Electronics Review* **14**, 279–308 (2012).
6. E.S. Barr, "Historical survey of the early development of the infrared spectral region", *American Journal of Physics* **28**, 42–54 (1960).
7. E.S. Barr, "The infrared pioneers—II. Macedonio Melloni", *Infrared Physics* **2**, 67–73 (1962).
8. E.S. Barr, "The infrared pioneers—III. Samuel Pierpont Langley", *Infrared Physics* **3**, 195–206 (1963).
9. W. Smith, "Effect of light on selenium during the passage of an electric current", *Nature* **7**, 303 (1873).
10. M.F. Doty, *Selenium, List of References, 1917–1925*, New York Public Library, New York, 1927.
11. J.C. Bose, Detector for electrical disturbances. U. S. Patent 755840, 1904.
12. T.W. Case, "Notes on the change of resistance of certain substrates in light", *Physical Review* **9**, 305–310 (1917).
13. T.W. Case, "The thalofide cell: A new photoelectric substance", *Physical Review* **15**, 289 (1920).
14. R.D. Hudson and J.W. Hudson, *Infrared Detectors*, Dowden, Hutchinson & Ross, Stroudsburg, 1975.
15. E.W. Kutzscher, "Review on detectors of infrared radiation", *Electro-Optical Systems Design* **5**, 30 (1973).
16. D.J. Lovell, "The development of lead salt detectors", *American Journal of Physics* **37**, 467–478 (1969).
17. R.J. Cushman, "Film-type infrared photoconductors", *Proceedings of the IRE* **47**, 1471–1475 (1959).
18. P.R. Norton, "Infrared detectors in the next millennium", *Proceedings of SPIE* **3698**, 652–665 (1999).
19. L. Hoddeson, "The discovery of the point-contact transistor". Historical Studies in the Physical Sciences. University of California Press 12(1), 41–76 (1981).
20. A. Rogalski, *Infrared and Terahertz Detectors*, CRC Press, Boca Raton, 2019.
21. E. Burstein, G. Pines, and N. Sclar, "Optical and photoconductive properties of silicon and germanium", in Photoconductivity Conference at Atlantic City, pp. 353–413, ed. R. Breckenbridge, B. Russell, and E. Hahn, Wiley, New York, 1956.

22. R.A. Soref, "Extrinsic IR photoconductivity of Si doped with B, Al, Ga, P, As or Sb", *Journal of Applied Physics* **38**, 5201–5209 (1967).
23. W.S. Boyle and G.E. Smith, "Charge-coupled semiconductor devices", *Bell Systems Technical Journal* **49**, 587–593 (1970).
24. F. Shepherd and A. Yang, "Silicon Schottky retinas for infrared imaging", *IEDM Technical Digest*, 310–313 (1973).
25. W.D. Lawson, S. Nielson, E.H. Putley, and A.S. Young, "Preparation and properties of HgTe and mixed crystals of HgTe-CdTe", *Journal of Physics and Chemistry of Solids* **9**, 325–329 (1959).
26. P.W. Kruse, M.D. Blue, J.H. Garfunkel, and W.D. Saur, "Long wavelength photoeffects in mercury selenide, mercury telluride and mercury telluride-cadmium telluride", *Infrared Physics* **2**, 53–60 (1962).
27. J. Melngailis and T.C. Harman, "Single-crystal lead-tin chalcogenides", in *Semiconductors and Semimetals*, Vol. 5, pp. 111–174, ed. R.K. Willardson and A.C. Beer, Academic Press, New York, 1970.
28. T.C. Harman and J. Melngailis, "Narrow gap semiconductors", in *Applied Solid State Science*, Vol. 4, pp. 1–94, ed. R. Wolfe, Academic Press, New York, 1974.
29. A. Rogalski and J. Piotrowski, "Intrinsic infrared detectors", *Progress in Quantum Electronics* **12**, 87–289 (1988).
30. S. Borrello and H. Levinstein, "Preparation and properties of mercury doped infrared detectors", *Journal of Applied Physics* **33**, 2947–2950 (1962).
31. C.T. Elliott, D. Day, and B.J. Wilson, "An integrating detector for serial scan thermal imaging", *Infrared Physics* **22**, 31–42 (1982).
32. A. Blackburn, M.V. Blackman, D.E. Charlton, W.A.E. Dunn, M.D. Jenner, K.J. Oliver, and J.T.M. Wotherspoon, "The practical realisation and performance of SPRITE detectors", *Infrared Physics* **22**, 57–64 (1982).
33. A.W. Hoffman, P.L. Love, and J.P. Rosbeck, "Mega-pixel detector arrays: Visible to 28 μm", *Proceedings of SPIE* **5167**, 194–203 (2004).
34. "Raytheon's LRAS3's sensor-to-missile capability gives warfighters remote, safe options", https://raytheon.mediaroom.com/2015-04-02-Raytheons-LRAS3s-sensor-to-missile-capability-gives-warfighters-remote-safe-options
35. O. Iwert and B. Delabrea, "The challenge of highly curved monolithic imaging detectors", *Proceedings of SPIE* **7742**, 774227-1–9 (2010).
36. K.-H. Jeong, J. Kim, and L.P. Lee, "Biologically inspired artificial compound eyes", *Science* **312**, 557–561 (2006).
37. Y.M. Song, Y. Xie, V. Malyarchuk, J. Xiao, I. Jung, K.-J. Choi, Z. Liu, H. Park, C. Lu, R.H. Kim, R. Li, K.B. Crozier, Y. Huang, and J.A Rogers, "Digital cameras with designs inspired by the arthropod eye", *Nature* **497**(7447), 95–99 (2013).
38. X. Tang, M.M. Ackerman, and P. Guyot-Sionnest, "Colloidal quantum dots based infrared electronic eyes for multispectral imaging", *Proceedings of SPIE* **11088**, 1108803-1–7 (2019).
39. Q. Lu, W. Liu, and X. Wang, "2-D material-based photodetectors on flexible substrates", in *Inorganic Flexible Optoelectronics: Materials and Applications*, pp. 117–142, ed. Z. Ma and D. Liu, Wiley-VCH Verlag, Berlin, 2019.

40. M.J.E. Golay, "A pneumatic infrared detector", *Review of Scientific Instruments* **18**, 357–362 (1947).

41. E.M. Wormser, "Properties of thermistor infrared detectors", *Journal of the Optical Society of America* **43**, 15–21 (1953).

42. R.W. Astheimer, "Thermistor infrared detectors", *Proceedings of SPIE* **443**, 95–109 (1983).

43. G.W. McDaniel and D.Z. Robinson, "Thermal imaging by means of the evaporograph", *Applied Optics* **1**, 311–324 (1962).

44. C. Hilsum and W.R. Harding, "The theory of thermal imaging, and its application to the absorption-edge image tube", *Infrared Physics* **1**, 67–93 (1961).

45. A.J. Goss, "The pyroelectric vidicon: A review", *Proceedings of SPIE* **807**, 25–32 (1987).

46. R.A. Wood and N.A. Foss, "Micromachined bolometer arrays achieve low-cost imaging", *Laser Focus World*, 101–106 (1993).

47. R.A. Wood, "Monolithic silicon microbolometer arrays", in *Semiconductors and Semimetals*, Vol. 47, pp. 45–121, ed. P.W. Kruse and D.D. Skatrud, Academic Press, San Diego, 1997.

48. C.M. Hanson, "Hybrid pyroelectric–ferroelectric bolometer arrays", in *Semiconductors and Semimetals*, Vol. 47, pp. 123–174, ed. P.W. Kruse and D.D. Skatrud, Academic Press, San Diego, 1997.

49. P.W. Kruse, "Uncooled IR focal plane arrays", *Opto-Electronics Review* 7, 253–258 (1999).

50. R.A. Wood, "Uncooled microbolometer infrared sensor arrays", in *Infrared Detectors and Emitters: Materials and Devices*, pp. 149–174, ed. P. Capper and C.T. Elliott, Kluwer Academic Publishers, Boston, 2000.

51. R.W. Whatmore and R. Watton, "Pyroelectric materials and devices", in *Infrared Detectors and Emitters: Materials and Devices*, pp. 99–147, ed. P. Capper and C.T. Elliott, Kluwer Academic Publishers, Boston, 2000.

52. P.W. Kruse, *Uncooled Thermal Imaging. Arrays*, Systems, and Applications, SPIE Press, Bellingham, 2001.

# CHAPTER 2

# Infrared Detector Characterization

O PTICAL RADIATION IS REGARDED as the radiation over the range from vacuum ultraviolet to the submillimeter wavelength (25 nm to 3000 μm). The terahertz (THz) region of the electromagnetic spectrum (Fig. 2.1) is often described as the final unexplored area of the spectrum and still presents a challenge for both electronic and photonic technologies. It is frequently treated as the spectral region within the frequency range $\nu \approx 0.1$–10 THz ($\lambda \approx 3$ mm–30 μm) and is partly overlapping with the loosely treated submillimeter (sub-mm) wavelength band $\nu \approx 0.1$–3 THz ($\lambda \approx 3$ mm–100 μm).

## 2.1 CLASSIFICATION OF INFRARED DETECTORS

The majority of optical detectors can be classified into two broad categories: photon detectors (also called quantum detectors) and thermal detectors.

### 2.1.1 Photon Detectors

In photon detectors, the radiation is absorbed within the material by interaction with electrons either bound to lattice atoms, to impurity atoms, or with free electrons. The observed electrical output signal results from the changed electronic energy distribution. The fundamental optical excitation processes in semiconductors are illustrated in Fig. 2.2. In quantum wells [Fig. 2.2(b)] the intersubband absorption takes place between the

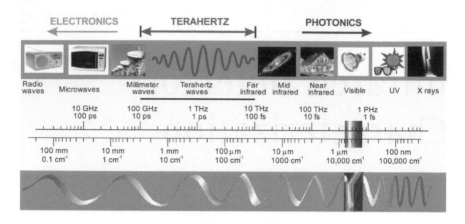

FIGURE 2.1 The electromagnetic spectrum (after Ref. [1]).

energy levels of a quantum well associated with the conduction band (n-doped) or the valence band (p-doped). In the case of a type-II InAs/GaSb superlattice [Fig. 2.2(c)], the superlattice bandgap is determined by the energy difference between the electron miniband $E_1$ and the first heavy-hole state $HH_1$ at the Brillouin zone center. A consequence of the type-II band alignment is spatial separation of electrons and holes.

Relative response of infrared detectors is plotted as a function of wavelength, with a vertical scale of either $W^{-1}$ or photon$^{-1}$ (Fig. 2.3). The photon detectors show a selective wavelength dependence of response per unit incident radiation power. Their response is proportional to the rate of arrival of photons, as the energy per photon is inversely proportional to the wavelength. In consequence, the spectral response increases linearly with increasing wavelength [Fig. 2.3(a)], until the cut-off wavelength is reached, which is determined by the detector material. The cut-off

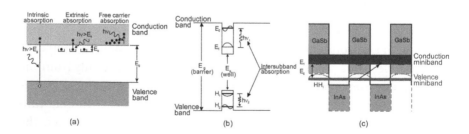

FIGURE 2.2 Optical excitation processes in (a) bulk semiconductors, (b) quantum wells, and (c) type-II InAs/GaSb superlattices.

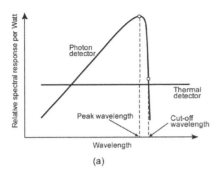 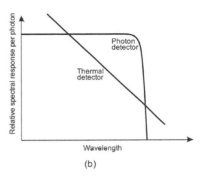

FIGURE 2.3   Relative spectral response for a photon and thermal detector for (a) constant incident radiant power and (b) photon flux, respectively.

wavelength is usually specified as the long-wavelength point at which the detector responsivity falls to 50% of the peak responsivity.

Thermal detectors tend to be spectrally flat in the first case (their response is proportional to the energy absorbed), so that they exhibit a flat spectral response [Fig 2.3(a)], whereas photon detectors are generally flat in the second case [Fig. 2.3(b)].

Photon detectors exhibit both good signal-to-noise performance and a very fast response. But, to achieve this, the photon IR detectors may require cryogenic cooling. This is necessary to prevent the thermal generation of charge carriers. The thermal transitions compete with the optical ones, making non-cooled devices very noisy.

Depending on the nature of the interaction, the class of photon detectors is further sub-divided into different types (Table 2.1). The most important are intrinsic detectors, extrinsic detectors, photoemissive detectors (Schottky barriers), and quantum well detectors [2]. Different types of detectors are briefly characterized in Table 2.2 [3].

There is a fundamental relationship between the temperature of the background viewed by the detector and the lower temperature at which the detector must operate to achieve background-limited performance (BLIP). HgCdTe photodetectors, with a cut-off wavelength of 12.4 μm, operate at 77 K. One can scale the results of this example to other temperatures and cut-off wavelengths by noting that, for a given level of detector performance, $T\lambda_c \approx$ constant [4]; i.e., the longer the $\lambda_c$, the lower $T$ is, when their product remains roughly constant. This relationship holds because quantities that determine detector performance vary mainly as an exponential of $E_{exc}/kT = hc/kT\lambda_c$, where $E_{exc}$ is the excitation energy, $k$ is Boltzmann's constant, $h$ is Planck's constant, and $c$ is the velocity of light.

TABLE 2.1 Comparison of Infrared Detectors

| Detector Type | | | Advantages | Disadvantages |
|---|---|---|---|---|
| Thermal (thermopile, bolometers, pyroelectric) | | | Light, rugged, reliable, and low cost; Room temperature operation | Low detectivity at high frequency; Slow response (ms order) |
| Photon | Intrinsic | IV–VI (PbS, PbSe, PbSnTe) | Easier to prepare; More stable materials | Very high thermal expansion coefficient; Large permittivity |
| | | II–VI (HgCdTe) | Easy bandgap tailoring; Well-developed theory and experiment.; Multicolor detectors | Non-uniformity over large area; High cost in growth and processing; Surface instability |
| | | III–V (InGaAs, InAs, InSb, InAsSb) | Good material and dopants; Advanced technology; Possible monolithic integration | Heteroepitaxy with large lattice mismatch; Long-wavelength cut-off limited to 7 μm for InAsSb at 77 K |
| | Extrinsic (Si:Ga, Si:As, Ge:Cu, Ge:Hg) | | Very-long-wavelength operation; Relatively simple technology | High thermal generation; Extremely-low-temperature operation |
| | Free carriers (PtSi, Pt$_2$Si, IrSi) | | Low cost, high yields; Large and close-packed 2-D arrays | Low quantum efficiency; Low-temperature operation |
| | Quantum wells | Type I (GaAs/AlGaAs, InGaAs/AlGaAs)) | Matured material growth; Good uniformity over large area; Multicolor detectors | High thermal generation; Complicated design and growth |
| | | Type II (InAs/GaSb, InAs/InAsSb) | Low Auger recombination rate; Easy wavelength control; Multicolor detectors | Complicated design and growth; Sensitive to the interfaces |
| | Quantum dots | InAs/GaAs, InGaAs/InGaP, Ge/Si | Normal incidence of light; Low thermal generation | Complicated design and growth |

TABLE 2.2 Photon Detectors

| Mode of Operation | Schematic of Detector | Operation and Properties |
|---|---|---|
| Photoconductor | 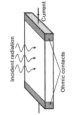 | This is essentially a radiation-sensitive resistor, generally a semiconductor, in either thin-film or bulk form. A photon may release an electron–hole pair or an impurity-bound charge carrier, thereby increasing the electrical conductivity. In almost all cases, the change in conductivity is measured by means of electrodes attached to the sample. For low-resistance material, the photoconductor is usually operated in a constant current circuit. For high-resistance photoconductors, a constant voltage circuit is preferred and the signal is detected as a change in current in the bias circuit. |
| Blocked-impurity-band (BIB) detector | 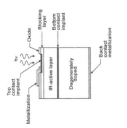 | The active region of a BIB detector structure, usually based on epitaxially grown n-type material, is sandwiched between a higher-doped degenerate substrate electrode and an undoped blocking layer. Doping of the active layer is high enough for the onset of an impurity band to display a high quantum efficiency for impurity ionization (in the case of Si:As BIB, the active layer is doped to $\approx 5 \times 10^{17}$ cm$^{-3}$). The device exhibits a diode-like characteristic, except that photoexcitation of electrons takes place between the donor impurity and the conduction band. The heavily doped n-type IR-active layer has a small concentration of negatively charged compensating acceptor impurities. In the absence of an applied bias, charge neutrality requires an equal concentration of ionized donors. Whereas the negative charges are fixed at acceptor sites, the positive charges associated with ionized donor sites (D$^+$ charges) are mobile and can propagate through the IR-active layer *via* the mechanism of hopping between occupied (D$^0$) and vacant (D$^+$) neighboring sites. A positive bias to the transparent contact creates a field that drives the pre-existing D$^+$ charges towards the substrate, whereas the undoped blocking layer prevents the injection of new D$^+$ charges. A region depleted of D$^+$ charges is therefore created, with a width depending on the applied bias and on the compensating acceptor concentration. |

*(Continued)*

TABLE 2.2 (CONTINUED)   Photon Detectors

| Mode of Operation | Schematic of Detector | Operation and Properties |
|---|---|---|
| p-n junction photodiode | 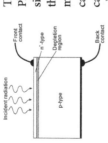 | This is the most widely used photovoltaic detector but is rather rarely used as a THz detector. Photons with energy greater than the energy gap create electron–hole pairs in the material on both sides of the junction. By diffusion, the electrons and holes generated within a diffusion length from the junction reach the space-charge region, where they are separated by the strong electric field; minority carriers become majority carriers on the other side. This way, a photocurrent is generated, causing a change in voltage across the open-circuit cell or a current to flow in the short-circuited case. The limiting noise level of photodiodes can ideally be $\sqrt{2}$ times lower than that of the photoconductor, due to the absence of recombination noise. Response times are generally limited by device capacitance and detector-circuit resistance. |
| nBn detector | | The nBn detector consists of a narrow-gap n-type absorber layer (AL), a thin wide-gap barrier layer (BL), and a narrow-gap n-type contact layer (CL). The thin wide-gap BL presents a large barrier in the conduction band that eliminates electron flow. Current through the nBn detector relies on transport of mobile holes through drift and diffusion in the BL between the two n-type narrow-gap regions. Effectively, the nBn detector is designed to reduce the dark current (generation–recombination current originating within the depletion layer) and noise, without impeding the photocurrent (signal). In particular, the barrier serves to reduce the surface leakage current. The nBn detector operates as a unipolar unity-gain detector and its design can be described as being a hybrid between a photoconductor and a photodiode. |

*(Continued)*

TABLE 2.2 (CONTINUED)   Photon Detectors

| Mode of Operation | Schematic of Detector | Operation and Properties |
| --- | --- | --- |
| Metal-insulator-semiconductor (MIS) photodiode | | The MIS device consists of a metal gate separated from a semiconductor surface by an insulator (I). By applying a negative voltage to the metal electrode, electrons are repelled from the I–S interface, creating a depletion region. When incident photons create hole–electron pairs, the minority carriers drift away to the depletion region and the volume of the depletion region shrinks. The total amount of charges that a photogate can collect is defined as its well capacity. The total well capacity is decided by the gate bias, the insulator thickness, the area of the electrodes, and the background doping of the semiconductor. Numerous such photogates with proper clocking sequence form a charge-coupled device (CCD) imaging array. |
| Schottky barrier photodiode | | Schottky barrier photodiodes reveal some advantages over p–n junction photodiodes: fabrication simplicity (deposition of metal barrier on n(p)-semiconductor), absence of high-temperature diffusion processes, and high speed of response. Since it is a majority carrier device, minority carrier storage and removal problems do not exist, and therefore higher bandwidths can be expected. The thermionic emission process in the Schottky barrier is much more efficient than the diffusion process and, therefore, for a given built-in voltage, the saturation current in a Schottky diode is several orders of magnitude higher than in the p–n junction. |

The detector temperature of low-background operation can be approximated as

$$T_{max} = \frac{300 \text{ K}}{\lambda_c \,[\mu m]}. \qquad (2.1)$$

The general trend is illustrated in Fig. 2.4 for six high-performance detector materials suitable for low-background applications: Si, InGaAs, InSb, HgCdTe photodiodes, and Si:X (X = As and Sb) blocked-impurity-band (BIB) detectors, and extrinsic Ge:Ga unstressed and stressed detectors. Terahertz photoconductors are operated in extrinsic mode.

The most widely used photovoltaic detector is the p-n junction, where a strong internal electric field exists across the junction, even in the absence of radiation. Photons incident on the junction produce free hole–electron pairs that are separated by the internal electric field across the junction, causing a change in voltage across the open-circuit cell or causing a current to flow in the short-circuited case. Due to the absence of recombination noise, the noise level of the limiting p–n junction can ideally be $\sqrt{2}$ times lower than that of the photoconductor.

Photoconductors that utilize excitation of an electron from the valence to the conduction band are called intrinsic detectors, whereas those which

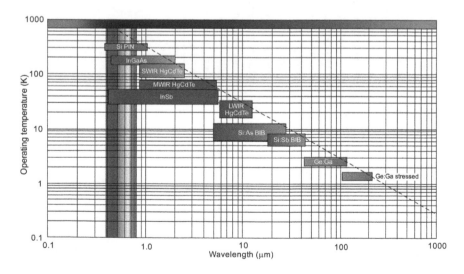

FIGURE 2.4 Operating temperatures for low-background material systems with their spectral band of greatest sensitivity. The dashed line indicates the trend toward lower operating temperature for longer-wavelength detection.

operate by exciting electrons into the conduction band or holes into the valence band from impurity states within the band (impurity-bound states in energy gap, quantum wells, or quantum dots), are called extrinsic detectors. A key difference between intrinsic and extrinsic detectors is that extrinsic detectors require considerable cooling to achieve high sensitivity at a given spectral response cut-off, in comparison with intrinsic detectors. Low-temperature operation is associated with longer-wavelength sensitivity to suppress noise due to thermally induced transitions between close-lying energy levels. Intrinsic detectors are most commonly used at the short wavelengths, below 20 μm. In the longer-wavelength region, the photoconductors are operated in extrinsic mode. One advantage of photoconductors is their current gain, which is equal to the recombination time divided by the majority carrier transit time. This current gain leads to higher responsivity than is possible with non-avalanching photovoltaic detectors. However, a series problem of photoconductors operated at low temperature is nonuniformity of the detector element due to recombination mechanisms at the electrical contacts and its dependence on electrical bias.

More recently, interfacial work-function internal photoemission detectors, quantum well, and quantum dot detectors, which can be included in extrinsic photoconductors, have been proposed, especially for IR and THz spectral bands [2,5]. The very fast time response of quantum-well semiconductor detectors and quantum dot semiconductor detectors makes them attractive for heterodyne detection.

## 2.1.2 Thermal Detectors

The second class of detectors is thermal detectors. In a thermal detector, shown schematically in Fig. 2.5, the incident radiation is absorbed to change the material temperature, and the resultant change in some physical property is used to generate an electrical output. The detector is suspended on lags, which are connected to the heat sink. The signal does not depend upon the photonic nature of the incident radiation. Thus, thermal effects are generally wavelength independent [Fig. 2.3(a)]; the signal depends upon the radiant power (or its rate of change) but not upon its spectral content. Since the radiation can be absorbed in a black surface coating, the spectral response can be very broad. Attention is directed toward three approaches that have found the greatest utility in infrared technology, namely bolometers, pyroelectric detectors, and thermoelectric effects. The thermopile is one of the oldest IR detectors, and is a collection of thermocouples, connected in series, to achieve greater temperature

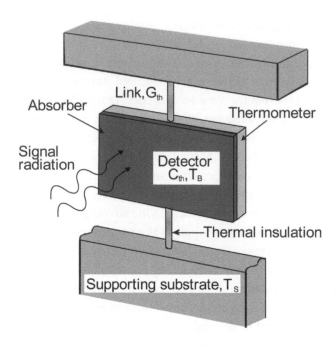

FIGURE 2.5   Schematic diagram of a thermal detector.

sensitivity. In pyroelectric detectors, a change in the internal electrical polarization is measured, whereas, in the case of thermistor bolometers, a change in the electrical resistance is measured. For a long time, thermopiles were slow, insensitive, bulky, and costly devices. But with developments in semiconductor technology, thermopiles can be optimized for specific applications. Recently, thanks to conventional complementary metal-oxide semiconductor (CMOS) processes, the on-chip circuitry technology of thermopiles has opened the door to mass production.

Usually, a bolometer is a thin, blackened flake or slab, the impedance of which is highly temperature dependent. Bolometers may be divided into several types. The types most commonly used are the metal, the thermistor, and the semiconductor bolometers. A fourth type is the superconducting bolometer. This bolometer operates on a conductivity transition, in which the resistance changes dramatically over the transition temperature range. Fig. 2.6 shows schematically the temperature dependence of resistance of the different types of bolometers.

Many types of thermal detectors are operated in the wide spectral range of electromagnetic radiation. The operation principles of thermal detectors are briefly described in Table 2.3.

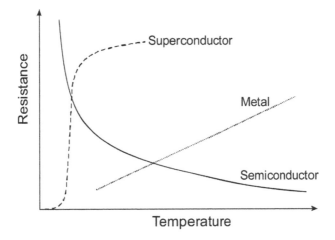

FIGURE 2.6 Temperature dependence of resistance of three bolometer material types.

The microbolometer detectors are now produced in larger volumes than all other IR array technologies combined. At present, $VO_x$ microbolometer arrays are clearly the most used technology for uncooled detectors. $VO_x$ is the winner in the battle between the amorphous silicon bolometers and the barium strontium titanate (BST) ferroelectric detectors.

The high cost of cryogenically cooled imagers, of around US$ 50,000, restricts their installation to critical military applications, involving operations in complete darkness. The commercial systems (microbolometer imagers, radiometers, and ferroelectric imagers) are derived from military systems that are too costly for widespread use. Imaging radiometers employ linear thermoelectric arrays operating in the snapshot mode; they are less costly than the TV-rate imaging radiometers that employ microbolometer arrays. As the volume of production increases, the cost of commercial systems will inevitably decrease. The current market price for a low-cost thermal imager is generally below US$1000. Recently, the first thermal imaging smartphone was launched [7].

## 2.2 DETECTOR FIGURES OF MERIT

It is difficult to measure the performance characteristics of infrared detectors because of the large number of experimental variables involved. A variety of environmental, electrical, and radiometric parameters must be considered and carefully controlled. With the advent of large, two-dimensional (2D) detector arrays, detector testing has become even more complex and demanding.

TABLE 2.3   Thermal Detectors

| Mode of Operation | Schematic of Detector | Operation and Properties |
|---|---|---|
| Thermopile | 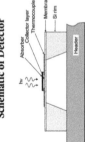 | The thermocouple is usually a thin, blackened flake, connected thermally to the junction of two dissimilar metals or semiconductors. Heat absorbed by the flake causes a temperature rise at the junction, and hence a thermoelectric electromotive force is developed that can be measured. Although thermopiles are not as sensitive as bolometers and pyroelectric detectors, they will replace these in many applications due to their reliable characteristics and good cost/performance ratio. Thermocouples are widely used in spectroscopy. |
| Bolometer Metal Semiconductor Superconductor Hot electron | | The bolometer is a resistive element constructed from a material with a very low thermal capacity and a large temperature coefficient, so that the absorbed radiation produces a large change in resistance. The change in resistance is similar to that in a photoconductor, although the basic detection mechanisms are different. In the case of a bolometer, radiant power produces heat within the material, which, in turn, produces the resistance change. There is no direct photon–electron interaction. Initially, most bolometers were the thermistor type, made from oxides of manganese, cobalt, or nickel. At present, microbolometers are fabricated in large format arrays for thermal imaging applications. Some extremely sensitive low-temperature semiconductor and superconductor bolometers are used in the THz region. |

*(Continued)*

TABLE 2.3 (CONTINUED)   Thermal Detectors

| Mode of Operation | Schematic of Detector | Operation and Properties |
|---|---|---|
| Pyroelectric detector | 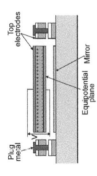 | The pyroelectric detector can be considered as a small capacitor, with two conducting electrodes mounted perpendicularly to the direction of spontaneous polarization. During incident radiation, the change in polarization appears as a charge on the capacitor and a current is generated, the magnitude of which depends on the temperature rise and the pyroelectrical coefficient of the material. The signal, however, must be chopped or modulated. The detector sensitivity is limited, either by amplifier noise or by loss-tangent noise. Response speed can be engineered, making pyroelectric detectors useful for fast laser pulse detection, although with a proportional decrease in sensitivity. |
| Golay cell | 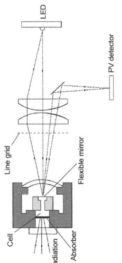 | The Golay cell consists of a hermetically sealed container filled with gas (usually xenon for its low thermal conductivity) and arranged so that expansion of the gas under heating by a photon signal distorts a flexible membrane on which a mirror is mounted. The movement of the mirror is used to deflect a beam of light shining on a photocell and so producing a change in the photocell current as the output. In modern Golay cells, the photocell is replaced by a solid-state photodiode, and a light-emitting diode is used for illumination.<br><br>The performance of the Golay cell is limited only by the temperature noise associated with the thermal exchange between the absorbing film and the detector gas; consequently the detector can be extremely sensitive, with $D^* \approx 3 \times 10^9$ cmHz$^{1/2}$W$^{-1}$, and responsivities of $10^5$ to $10^6$ V/W. The response time is quite long, typically 15 msec. |

*(Continued)*

TABLE 2.3 (CONTINUED)   Thermal Detectors

| Mode of Operation | Schematic of Detector | Operation and Properties |
| --- | --- | --- |
| Thermomechanical [6] | Infrared radiation<br><br>Substrate<br><br>Δz | The thermomechanical detector consists of coated microcantilevers, with a metal as the sensing active layer to form the bimaterial. The microcantilever is attached mechanically and electrically to the substrate at the end by an anchor, and the second end is free to bend under the influence of any changes in stress along the arm. IR radiation is absorbed by the microcantilever paddle materials, along with a tuned resonant absorption cavity. The absorbed radiation is converted to heat in the microcantilever structure, thermally isolated from the substrate by thermal isolation arms like those used in bolometers. The thermomechanical detectors can be combined with a number of different readout techniques. Depending on the readout technique, they can be devoted to capacitive, optical, piezoresistive, and electron tunneling. |

This section is intended to serve as an introductory reference for the testing of infrared detectors. Numerous texts and papers cover this topic, including: *Infrared System Engineering* [8] by R.D. Hudson; *The Infrared Handbook* [9], edited by W.L. Wolfe and G.J. Zissis; *The Infrared and Electro-Optical Systems Handbook* [10], edited by W.D. Rogatto; and *Fundamentals of Infrared Detector Operation and Testing* [11] by J.D. Vincent, and the second edition, Vincent et al. (2016), of the latter book [12]. In this chapter, we have restricted our consideration to detectors, the output of which consists of an electrical signal that is proportional to the radiant signal power.

## 2.2.1 Responsivity

The responsivity of an infrared detector is defined as the ratio of the root-mean-square (rms) value of the fundamental component of the electrical output signal of the detector to the rms value of the fundamental component of the input radiation power. The units of responsivity are volts per watt (V/W) or amperes per watt (A/W).

The voltage (or analogous current) spectral responsivity is given by

$$R_v(\lambda, f) = \frac{V_s}{\Phi_e(\lambda)\Delta\lambda},$$  (2.2)

where $V_s$ is the signal voltage due to $\Phi_e$, and $\Phi_e(\lambda)$ is the spectral radiant incident power (in W).

An alternative to the above monochromatic quality, the blackbody responsivity, is defined by the equation

$$R_v(T, f) = \frac{V_s}{\int_0^\infty \Phi_e(\lambda)\Delta\lambda},$$  (2.3)

where the incident radiant power is the integral over all wavelengths of the spectral density of power distribution $\Phi_e(\lambda)$ from a blackbody. The responsivity is usually a function of the bias voltage, the operating electrical frequency, and the wavelength.

## 2.2.2 Noise Equivalent Power

The noise equivalent power (*NEP*) is the incident power on the detector, generating a signal output equal to the rms noise output. Stated in another way, the *NEP* is the signal level that produces a signal-to-noise ratio (*SNR*) of 1. It can be written in terms of responsivity:

$$NEP = \frac{V_n}{R_v} = \frac{I_n}{R_i}.$$ (2.4)

The unit of *NEP* is watts.

The *NEP* is also quoted for a fixed reference bandwidth, which is often assumed to be 1 Hz. This "*NEP* per unit bandwidth" has a unit of watts per square root hertz (W/Hz$^{1/2}$).

### 2.2.3  Detectivity

The detectivity *D* is the reciprocal of *NEP*:

$$D = \frac{1}{NEP}.$$ (2.5)

It was found by Jones [13], that, for many detectors, the *NEP* is proportional to the square root of the detector signal, which is proportional to the detector area, $A_d$. This means that both *NEP* and detectivity are functions of electrical bandwidth and detector area, so a normalized detectivity *D*\* (or *D*-star), suggested by Jones [13], is defined as:

$$D^* = D\left(A_d \Delta f\right)^{1/2} = \frac{\left(A_d \Delta f\right)^{1/2}}{NEP}.$$ (2.6)

The importance of *D*\* is that this figure of merit permits comparison of detectors of the same type, but having different areas. Either a spectral or blackbody *D*\* can be defined in terms of the corresponding type of *NEP*.

Useful expressions equivalent to Eq. (2.6) include:

$$D^* = \frac{\left(A_d \Delta f\right)^{1/2}}{V_n} R_v = \frac{\left(A_d \Delta f\right)^{1/2}}{I_n} R_i = \frac{\left(A_d \Delta f\right)^{1/2}}{\Phi_e}(SNR),$$ (2.7)

where *D*\* is defined as the rms signal to noise ratio (*SNR*) in a 1-Hz bandwidth per unit rms incident radiant power per square root of detector area. *D*\* is expressed in unit cmHz$^{1/2}$W$^{-1}$, a unit which recently has been referred to as a "Jones".

Spectral detectivity curves for several commercially available IR and THz detectors are shown in Fig. 2.7. Interest has focused mainly on the

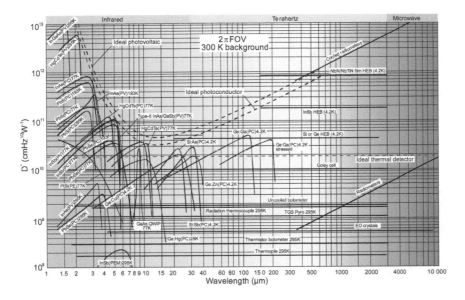

FIGURE 2.7 Comparison of the $D^*$ of various available detectors when operated at the indicated temperature. Chopping frequency is 1000 Hz for all detectors except for 10 Hz for the thermopile, thermocouple, thermistor bolometer, Golay cell, and pyroelectric detector. Each detector is assumed to view a surrounding hemisphere ($2\pi$ field of view) at a temperature of 300 K. Theoretical curves for the background-limited $D^*$ (dashed lines) for ideal photovoltaic and photoconductive detectors and thermal detectors are also shown. PC: photoconductive detector; PV: photovoltaic detector; PEM: photoelectromagnetic detector; and HEB: hot electron bolometer (after Ref. [14]).

two atmospheric windows 3–5 μm (medium-wavelength IR, MWIR) and 8–14 μm (long-wavelength IR, LWIR); atmospheric transmission is the highest in these bands and the emissivity maximum of the objects at $T \approx 300$ K is at the wavelength $\lambda \approx 10$ μm. In recent years, there has been increasing interest in longer wavelengths, stimulated by space and tera-hertz applications. The spectral character of the background is influenced by the transmission of the atmosphere that controls the spectral ranges of the infrared for which the detector may be used when operating in the atmosphere.

## 2.2.4 Quantum Efficiency

A signal whose photon energy is sufficient to generate photocarriers will continuously lose energy as the optical field propagates through the semi-conductor. Inside the semiconductor, the field decays exponentially as energy is transferred to the photocarriers. The material can be characterized by an absorption length $\alpha$ and a penetration depth $1/\alpha$. Penetration depth is the point at which $1/e$ of the optical signal power remains.

The power absorbed in the semiconductor as a function of position within the material is then:

$$P_a = P_i\left(1-r\right)\left(1-e^{-\alpha x}\right). \tag{2.8}$$

The number of photons absorbed is the power (in W) divided by the photon energy ($E = h\nu$). If each absorbed photon generates a photocarrier, the number of photocarriers generated per number of incident photons for a specific semiconductor, with reflectivity $r$, is given by:

$$\eta\left(x\right) = \left(1-r\right)\left(1-e^{-\alpha x}\right), \tag{2.9}$$

where $0 \leq \eta \leq 1$ is a definition of the detector's quantum efficiency, as the number of electron–hole pairs generated per incident photon.

Figure 2.8 shows the quantum efficiency of some of the detector materials used to fabricate arrays of ultraviolet (UV), visible, and infrared detectors. Photocathodes and aluminum gallium nitride (AlGaN) detectors in the UV region are being developed. Silicon p-i-n diodes are shown with and without antireflection coating. Lead salts (PbS and PbSe) have

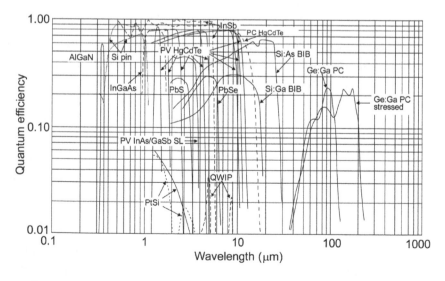

FIGURE 2.8   Quantum efficiency of different detectors (after Ref. [3]).

intermediate quantum efficiencies, whereas PtSi Schottky barrier types and quantum-well infrared photodetectors (QWIPs) have low values. InSb can respond from the near-UV out to 5.5 µm at 80 K. A suitable detector material for the near-IR (1.0–1.7 µm) spectral range is InGaAs lattice, matched to the InP. Various HgCdTe alloys, in both photovoltaic and photoconductive configurations, cover from 0.7 µm to over 20 µm. InAs/GaSb strained-layer superlattices have emerged as an alternative to the HgCdTe. Impurity-doped (Sb, As, and Ga) silicon BIB detectors operating at 10 K have a spectral response cut-off in the range of 16 to 30 µm. Impurity-doped Ge stressed detectors can extend the response out to 100–200 µm.

There are different methods of light coupling in a photodetector to enhance quantum efficiency [15]. A notable example of a method described for thin-film solar cells [16,17] can be applied to infrared photodetectors. In general, these absorption enhancement methods can be divided into four categories that use either optical concentration, antireflection structures, optical path increase, or light localization, as shown in Fig. 2.9. They are briefly described in Chapter 8 of monograph *Antimonide-based Infrared Detectors: A New Perspective* [3].

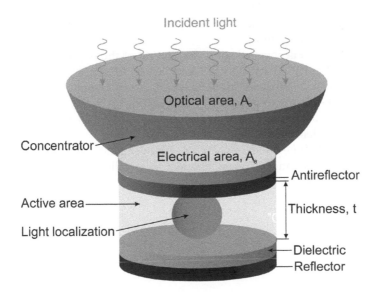

FIGURE 2.9 Different methods of absorption enhancement in a photodetector use of an optical concentrator, an antireflection structure, structures for optical path increase (cavity enhancement), and light localization structures.

## REFERENCES

1. A. Rogalski and F. Sizov, "Terahertz detectors and focal plane arrays", *Opto-Electronics Review* **19**(3), 346–404 (2011).
2. A. Rogalski, *Infrared and Terahertz Detectors*, 3rd edition, CRC Press, Boca Raton, 2019.
3. A. Rogalski, M. Kopytko, and P. Martyniuk, *Antimonide-based Infrared Detectors: A New Perspective*, SPIE Press, Bellingham, 2018.
4. D. Long, "Photovoltaic and photoconductive infrared detectors", in *Optical and Infrared Detectors*, pp. 101–147, ed. R.J. Keyes, Springer, Berlin, 1980.
5. A. Rogalski, "Quantum well photoconductors in infrared detector technology", *Journal of Applied Physics* **93**, 4355 (2003).
6. M. Steffanson and I.W. Rangelow, "Microthermomechanical infrared sensors", *Opto-Electronics Review* **22**, 1–15 (2014).
7. "The World's first thermal imaging smartphone", http://www.flir.com/home/news/details/?ID=74197
8. R.D. Hudson, *Infrared System Engineering*, Wiley, New York, 1969.
9. W.I. Wolfe and G.J. Zissis (eds.), *The Infrared Handbook*, Office of Naval Research, Washington, 1985.
10. W.D. Rogatto (ed.), *The Infrared and Electro-Optical Systems Handbook*, Infrared Information Analysis Center, Ann Arbor and SPIE Press, Bellingham, Washington, 1993.
11. J.D. Vincent, *Fundamentals of Infrared Detector Operation and Testing*, Wiley, New York, 1990.
12. J.D. Vincent, S.E. Hodges, J. Vampola, M. Stegall, and G. Pierce, *Fundamentals of Infrared and Visible Detector Operation and Testing*, Wiley, Hoboken, 2016.
13. R.C. Jones, "Phenomenological description of the response and detecting ability of radiation detectors", *Proceedings of IRE* **47**, 1495–1502 (1959).
14. A. Rogalski, "Progress in focal plane array technology", *Progress in Quantum Electronics* **36**(2/3), 342–473 (2012).
15. Z. Jakšić, *Micro and Nanophotonics for Semiconductor Infrared Detectors*, Springer, Heidelberg, 2014.
16. S.J. Fonash, *Solar Cell Device Physics*, Elsevier, Amsterdam, 2010.
17. G. Li, R. Zhu, and Y. Yang, "Polymer solar cells", *Nature Photonics* **6**(3), 153–161 (2012).

# CHAPTER 3

# Fundamental Detector Performance Limits

THIS CHAPTER DISCUSSES THE fundamental limitations to IR detector performance imposed by the statistical nature of the generation-recombination processes, and by radiometric considerations. We will try to establish the ultimate theoretical sensitivity limit that can be expected for a detector operating at a given temperature.

Photon detectors are fundamentally limited by generation-recombination noise arising from photon exchange with a radiation background. Thermal detectors are fundamentally limited by temperature fluctuation noise arising from radiant power exchange with a radiating background. Due to fundamentally different types of noise, these two classes of detectors have different dependencies of detectivities on wavelength and temperature. The photon detectors are favored at a long wavelength infrared spectral range and at lower operating temperatures. The thermal detectors are favored at a very long wavelength spectral range.

In this chapter, we first examine fundamental infrared detection processes for both categories of detectors. To start with, comparative studies of thermal and photon detectors are carried out.

## 3.1 PHOTON DETECTORS

In general, the photon detector can be considered as a slab of homogeneous material with actual "electrical" area, $A_e$, thickness, $t$, and volume, $A_e t$ (see Fig. 2.9). Usually, the optical and electrical areas of the device are the same or similar. However, the use of some kind of optical concentrator can increase the $A_o/A_e$ ratio by a large factor.

The detectivity, $D^*$, of an infrared photodetector is limited by generation and recombination rates $G$ and $R$ (in $m^{-6}s^{-1}$) in the active region of the device [1,2]. It can be expressed as:

$$D^* = \frac{\lambda}{2^{1/2}hc(G+R)^{1/2}}\left(\frac{A_o}{A_e}\right)^{1/2}\frac{\eta}{t^{1/2}},\qquad(3.1)$$

where $\lambda$ is the wavelength, $h$ is Planck's constant, $c$ is the velocity of light, and $\eta$ is the quantum efficiency.

For a given wavelength and operating temperature, the highest performance can be obtained by maximizing the ratio of the quantum efficiency to the square root of the sum of the sheet thermal generation and recombination rates $\eta/[(G+R)t]^{1/2}$. This means that high quantum efficiency must be obtained with a thin device.

A possible way to improve the performance of IR detectors is to reduce the physical volume of the semiconductor, thus reducing the amount of thermal generation. However, this must be achieved without a decrease in the quantum efficiency, optical area, or field of view (FOV) of the detector.

At equilibrium, the generation and recombination rates are equal. If we further assume $A_e = A_o$, the detectivity of an optimized infrared photodetector is limited by thermal processes in the active region of the device. It can be expressed as:

$$D^* = 0.31\frac{\lambda}{hc}k\left(\frac{\alpha}{G}\right)^{-1/2},\qquad(3.2)$$

where $1 \leq k \leq 2$ and is dependent on the contribution of recombination and backside reflection. The $k$-coefficient can be modified by using more sophisticated coupling of the detector with IR radiation, e.g., using photonic crystals or surface plasmon-polaritons.

The ratio of the absorption coefficient to the thermal generation rate, $\alpha/G$, is the fundamental figure of merit of any material intended for infrared photodetection. The $\alpha/G$ ratio *versus* temperature relationship for various material systems capable of bandgap tuning is shown in Fig. 3.1 for a hypothetical energy gap equal to 0.25 eV ($\lambda = 5$ μm) [Fig. 3.1(a)] and 0.124 eV ($\lambda = 10$ μm) [Fig. 3.1(b)]. Procedures used in calculations of $\alpha/G$ for different material systems are given in [3]. Analysis shows that

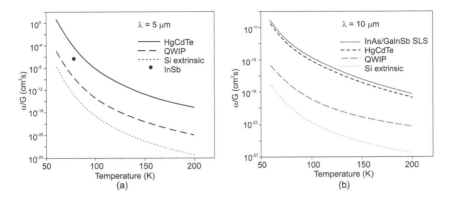

FIGURE 3.1 $\alpha/G$ ratio *versus* temperature for (a) MWIR ($\lambda = 5\ \mu$m) and (b) LWIR ($\lambda = 10\ \mu$m) photodetectors based on HgCdTe, QWIP, Si extrinsic, and type-II superlattice (for LWIR only) material technology (adapted after Ref. [4]).

narrow-bandgap semiconductors are more suitable for high-temperature photodetectors than competing technologies, such as extrinsic devices, QWIP (quantum well IR photodetector), and QDIP (quantum dot IR photodetector) devices. The main reason for the high performance of intrinsic photodetectors is the high density of states in the valence and conduction bands, which results in strong absorption of infrared radiation. Figure 3.1(b) predicts that the emerging competing IR material, the type-II superlattice, is the most efficient material technology for IR detection in the long-wavelength region, theoretically perhaps even better than HgCdTe if the influence of the Shockley–Read–Hall lifetime is not considered. It is characterized by a high absorption coefficient and relatively low fundamental (band-to-band) thermal generation rate. However, this theoretical prediction has not been confirmed by experimental data. It is also worth noting that, theoretically, AlGaAs/GaAs QWIP is also a better material than extrinsic silicon.

The ultimate performance of infrared detectors is reached when the detector and amplifier noise are low, compared with the photon noise. The photon noise is fundamental in the sense that it arises, not from any imperfection in the detector or its associated electronics, but rather from the detection process itself, as a result of the discrete nature of the radiation field. The radiation falling on the detector is a composite of that from the target and that from the background. The practical operating limit for most infrared detectors is not the signal fluctuation limit but the background fluctuation limit, also known as the background-limited infrared photodetector (BLIP) limit.

The expression for shot noise can be used to derive the BLIP detectivity:

$$D^*_{BLIP}(\lambda,T) = \frac{\lambda}{hc}k\left(\frac{\eta}{2\Phi_B}\right)^{1/2},\qquad(3.3)$$

where $\eta$ is the quantum efficiency, and $\Phi_B$ is the total background photon flux density reaching the detector, denoted as:

$$\Phi_B = sin^2(\theta/2)\int_0^{\lambda_c}\Phi(\lambda,T_B)d\lambda,\qquad(3.4)$$

where $\theta$ is the detector field-of-view angle.

Planck's photon emittance (in units of photons $cm^{-2}\ s^{-1}\ \mu m^{-1}$) at temperature $T_B$ is given by:

$$\Phi(\lambda,T_B) = \frac{2\pi c}{\lambda^4\left[exp(hc/\lambda kT_B)-1\right]} = \frac{1.885\times10^{23}}{\lambda^4\left[exp(14.388/\lambda kT_B)-1\right]}.\qquad(3.5)$$

Equation (3.3) holds for photovoltaic detectors, which are shot noise limited. Photoconductive detectors that are generation-recombination noise limited have a $D^*_{BLIP}$ lower by a factor of $2^{1/2}$:

$$D^*_{BLIP}(\lambda,T) = \frac{\lambda}{2hc}k\left(\frac{\eta}{\Phi_B}\right)^{1/2}.\qquad(3.6)$$

Once background-limited performance is reached, quantum efficiency $\eta$ is the only detector parameter that can influence a detector's performance.

Figure 3.2 shows the peak spectral detectivity of a background-limited photodetector, operating at 300, 230, and 200 K, *versus* the wavelength calculated for 300 K background radiation and hemispherical field-of-view (FOV) ($\theta=90$ deg). The minimum $D^*_{BLIP}$ (300 K) occurs at 14 μm and is equal to $4.6\times10^{10}$ cmHz$^{1/2}$/W. For some photodetectors, which operate under near-equilibrium conditions, such as non-sweep-out photoconductors, the recombination rate is equal to the generation rate. For these detectors, the contribution of recombination to the noise will reduce $D^*_{BLIP}$ by a factor of $2^{1/2}$. Note that $D^*_{BLIP}$ does not depend on area or the $A_o/A_e$ ratio. As a consequence, the background-limited performance cannot be improved by making $A_o/A_e$ large.

The highest performance possible will be obtained by the ideal detector with unity quantum efficiency and ideal spectral responsivity; $R(\lambda)$

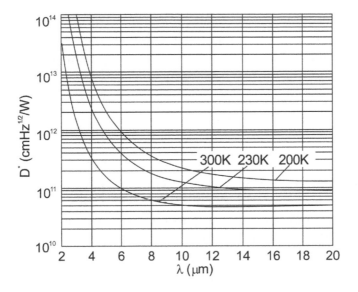

FIGURE 3.2 Calculated spectral detectivities of a photodetector limited by the hemispherical FOV background radiation of 300 K as a function of the peak wavelength for detector operating temperatures of 300, 230, and 200 K (after Ref. [2]).

increases with increasing wavelength to the cutoff wavelength $\lambda_c$, at which point the responsivity drops to zero. This limiting performance is of interest for comparison with actual detectors.

The detectivity of BLIP detectors can be improved by reducing the background photon flux, $\Phi_b$. Practically, there are two ways to do this: a cooled or reflective spectral filter to limit the spectral band, or a cooled shield to limit the angular field of view of the detector (as described above). The former eliminates background radiation from spectral regions in which the detector need not respond. The best detectors yield background-limited detectivities in quite narrow fields of view.

## 3.2 THERMAL DETECTORS

Thermal detectors operate on a simple principle, that, when heated by incoming radiation, their temperature increases and the temperature change is measured by any temperature-dependent mechanism, such as thermoelectric voltage, resistance, or pyroelectric voltage.

The simplest representation of the thermal detector is shown in Fig. 2.5. The detector is represented by a thermal capacitance, $C_{th}$, coupled *via* a thermal conductance, $G_{th}$, to a heat sink at a constant temperature, $T$.

In the absence of a radiation input, the average temperature of the detector will also be $T$, although it will exhibit a fluctuation near this value. When a radiation flux, $\Phi_o$, is received by the detector, the rise in temperature, $\Delta T$, is found by solving the heat balance equation [5,6] as:

$$\Delta T = \frac{\varepsilon \Phi_o}{\left(G_{th}^2 + \omega^2 C_{th}^2\right)^{1/2}}. \tag{3.7}$$

Equation (3.7) illustrates several features of a thermal detector. Clearly, it is advantageous to make $\Delta T$ as large as possible. To do this, the thermal capacity of the detector ($C_{th}$) and its thermal coupling to its surroundings ($G_{th}$) need to be as small as possible. The interaction of the thermal detector with the incident radiation needs to be optimized, while reducing as far as possible all other thermal contacts with its surroundings. This means that a small detector mass and fine connecting wires to the heat sink are desirable.

Equation (3.7) shows that, as $\omega$ is increased, the term $\omega^2 C_{th}^2$ will eventually exceed $G_{th}^2$ and then $\Delta T$ will fall inversely as $\omega$. A characteristic thermal response time for the detector can therefore be defined as

$$\tau_{th} = \frac{C_{th}}{G_{th}} = C_{th} R_{th}, \tag{3.8}$$

where the thermal resistance $R_{th} = 1/G_{th}$. Then Equation (3.7) can be written as

$$\Delta T = \frac{\varepsilon \Phi_o R_{th}}{\left(1 + \omega^2 \tau_{th}^2\right)^{1/2}}. \tag{3.9}$$

A typical value of the thermal time constant is in the millisecond range. This is much longer than the typical time constant of a photon detector. There is a trade-off between sensitivity, $\Delta T$, and frequency response. If one wants a high sensitivity, then a low frequency response is imposed upon the detector.

For further discussion, we introduce the coefficient, $K$, which reflects how well the temperature changes translate into the electrical output voltage of the detector:

$$K = \frac{\Delta V}{\Delta T}. \tag{3.10}$$

Then, the corresponding root-mean-square (rms) voltage signal, due to temperature changes, $\Delta T$, is:

$$\Delta V = K\Delta T = \frac{K\varepsilon\Phi_o R_{th}}{\left(1+\omega^2\tau_{th}^2\right)^{1/2}}. \tag{3.11}$$

The voltage responsivity, $R_v$, of the detector is the ratio of the output signal voltage, $\Delta V$, to the input radiation power, and is given by:

$$R_v = \frac{K\varepsilon R_{th}}{\left(1+\omega^2\tau_{th}^2\right)^{1/2}}. \tag{3.12}$$

As the last expression, Equation (3.12), shows, the low frequency voltage responsivity ($\omega \ll 1/\tau_{th}$) is proportional to the thermal resistance and does not depend on the heat capacitance. The opposite is true for high frequencies ($\omega \gg 1/\tau_{th}$). In this case, $R_v$ is not dependent on $R_{th}$ and is inversely proportional to the heat capacitance.

As stated previously, the thermal conductance from the detector to the outside world should be small (and hence the thermal resistance should be high). The lowest possible thermal conductance would occur when the detector is completely isolated from the environment under vacuum, with only radiative heat exchange between it and its heat-sink enclosure. Such an ideal model can give us the ultimate performance limit of a thermal detector. This limiting value can be estimated from the Stefan-Boltzmann total radiation law.

If the thermal detector has a receiving area, $A$, of emissivity, $\varepsilon$, when it is in thermal equilibrium with its surroundings, it will radiate a total flux, $A\varepsilon\sigma T^4$, where $\sigma$ is the Stefan-Boltzmann constant. Now, if the temperature of the detector is increased by a small amount, $dT$, the flux radiated, is increased by $4A\varepsilon\sigma T^3 dT$. Hence, the radiative component of the thermal conductance is:

$$G_R = \frac{1}{\left(R_{th}\right)_R} = \frac{d}{dT}\left(A\varepsilon\sigma T^4\right) = 4A\varepsilon\sigma T^3. \tag{3.13}$$

In this case

$$R_v = \frac{K}{4\sigma T^3 A\left(1+\omega^2\tau_{th}^2\right)^{1/2}}. \tag{3.14}$$

When the detector is in thermal equilibrium with the heat sink, the fluctuation in the power flowing through the thermal conductance into the detector [5,7] is:

$$\Delta P_{th} = \left(4KT^2G\right)^{1/2},$$ (3.15)

which will be smallest when $G$ assumes its minimum value (i.e., $G_R$). Then, $\Delta P_{th}$ will be at a minimum and its value gives the minimum detectable power for an ideal thermal detector.

The minimum detectable signal power – or noise equivalent power (*NEP*) – is defined as the rms signal power incident upon the detector required to equal the rms thermal noise power. Hence, if the temperature fluctuation associated with $G_R$ is the only source of noise:

$$\varepsilon NEP = \Delta P_{th} = \left(16A\varepsilon\sigma kT^5\right)^{1/2}$$ (3.16)

or

$$NEP = \left(\frac{16A\sigma kT^5}{\varepsilon}\right)^{1/2}.$$ (3.17)

If all the incident radiation is absorbed by the detector, $\varepsilon = 1$, and then

$$NEP = \left(16A\sigma kT^5\right)^{1/2} = 5.0 \times 10^{-11}\,\text{W};$$ (3.18)

for $A = 1$ cm², $T = 290$ K, and $\Delta f = 1$ Hz. For a small detector, with, e.g., $A = 10 \times 10$ µm², $NEP = 5.0 \times 10^{-14}$ W, and is three orders of magnitude lower than previously.

To determine the detectivity ($D^*$) of a detector, it is necessary to define a noise mechanism. For any detector, there are several noise sources that impose fundamental limits to the detection sensitivity.

One major noise is the Johnson noise, although two other fundamental noise sources are important for assessing the ultimate performance of a detector, namely thermal fluctuation noise and background fluctuation noise.

Thermal fluctuation noise arises from temperature fluctuations in the detector. These fluctuations are caused by heat conductance variations between the detector and the surrounding substrate with which the detector element is in thermal contact.

The variance in temperature ("temperature" noise) [6,7] can be shown to be:

$$\overline{\Delta T^2} = \frac{4kT^2 \Delta f}{1+\omega^2 \tau_{th}^2} R_{th} \qquad (3.19)$$

This equation shows that thermal conductance, $G_{th} = 1/R_{th}$, as the principal heat loss mechanism, is the key design parameter that affects the temperature fluctuation noise. The spectral noise voltage due to temperature fluctuations is:

$$V_{th}^2 = K^2 \overline{\Delta T^2} = \frac{4kT^2 \Delta f}{1+\omega^2 \tau_{th}^2} K^2 R_{th}. \qquad (3.20)$$

The third fundamental noise source is background noise resulting from radiative heat exchange between the detector at temperature $T_d$ and the surrounding environment at temperature $T_b$. It represents the ultimate limit of a detector's performance capability and is given, for a $2\pi$ field-of-view (FOV) [6,7], by:

$$V_b^2 = \frac{8k\varepsilon\sigma A \left(T_d^2 + T_b^2\right)}{1+\omega^2 \tau_{th}^2} K^2 R_{th}^2, \qquad (3.21)$$

where $\sigma$ is the Stefan-Boltzmann constant.

In addition to the fundamental noise sources mentioned above, $1/f$ is an additional noise source that is often found in the thermal detector and can affect detector performance.

The detectivity of a thermal detector is given by:

$$D^* = \frac{K\varepsilon R_{th} A^{1/2}}{\left(1+\omega^2 \tau_{th}^2\right)^{1/2} \left(\dfrac{4kT_d^2 K^2 R_{th}}{1+\omega^2 \tau_{th}^2} + 4kTR + V_{1/f}^2\right)^{1/2}}. \qquad (3.22)$$

In the case of a typical operation condition of the thermal detector, where it operates in a vacuum or in a gas environment at reduced pressures, heat conduction through the supporting microstructure of the device is a dominant heat loss mechanism. However, in the case of an extremely good thermal isolation, the principal heat loss mechanism can be reduced to only radiative heat exchange between the detector and its surroundings. In an atmospheric environment, heat conduction through air is likely to be the dominant heat dissipation mechanism.

The fundamental limit to the sensitivity of any thermal detector is set by the temperature fluctuation noise. Under these conditions at low frequencies ($\omega \ll 1/\tau_{th}$), from Equation (3.22), the limit results in:

$$D_{th}^* = \left( \frac{\varepsilon^2 A}{4kT_d^2 G_{th}} \right)^{1/2} .$$ 

(3.23)

It is assumed here that $\varepsilon$ is independent of wavelength, so that the spectral $D_\lambda^*$ and the blackbody $D^*(T)$ values are identical.

Figure 3.3 shows the dependence of detectivity on temperature and thermal conductance plotted for different detector active areas. It is clearly shown that improved performance of thermal detectors can be achieved by increasing thermal isolation between the detector and its surroundings.

If radiant power exchange is the dominant heat exchange mechanism, then $G$ is the first derivative, with respect to temperature, of the Stefan-Boltzmann function. In that case, known as the background fluctuation noise limit, from Equations (2.7) and (3.21), we have:

$$D_b^* = \left[ \frac{\varepsilon}{8k\sigma \left( T_d^5 + T_b^5 \right)} \right]^{1/2} .$$ 

(3.24)

Note that $D_b^*$ is independent of $A$, as is to be expected.

Figure 3.4 shows the photon noise-limited detectivity for an ideal thermal detector having an emissivity of unity, operated at 290 K and lower, as a function of background temperature. In many practical instances, the

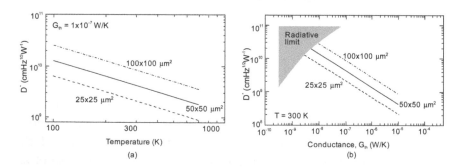

FIGURE 3.3 Temperature fluctuation noise-limited detectivity for thermal infrared detectors of different areas plotted (a) as a function of the detector temperature, and (b) as a function of the total thermal conductance between the detector and its surroundings (adapted after Ref. [8]).

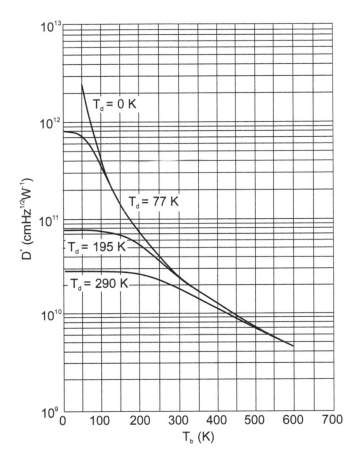

FIGURE 3.4  Temperature fluctuation noise-limited detectivity of thermal detectors as a function of detector temperatures $T_d$ and background temperature $T_b$, for $2\pi$ FOV and $\varepsilon = 1$.

temperature of the background, $T_b$, is room temperature, 290 K. The next figure (Fig. 3.5) shows the background-limited detectivity, as a function of the sensor temperature, for an ideal thermal detector.

Equations (3.23) and (3.24) and Fig. 3.4 assume that background radiation falls upon the detector from all directions, when the detector and background temperatures are equal, and from the forward hemisphere only when the detector is at cryogenic temperatures. We see that the highest possible $D^*$ to be expected for a thermal detector operated at room temperature and viewing a background at room temperature is $1.98 \times 10^{10}$ $\mathrm{cmHz^{1/2}W^{-1}}$. Even if the detector or background (but not both) were cooled to absolute zero, the detectivity would improve only by the square root of

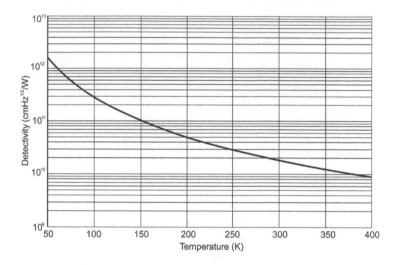

FIGURE 3.5    BLIP detectivity of a thermal detector as a function of the sensor temperature for 2 FOV and $\varepsilon = 1$.

two. This is the basic limitation of all thermal detectors. The background noise-limited photon detectors have higher detectivities because of their limited spectral responses (as shown in Fig. 2.7).

Most thermal detectors can be tailored to have somewhat different properties and the user should contact the manufacturer for detailed information. Table 3.1 gives the general flavor of the performance of different thermal detectors.

## 3.3  THE ULTIMATE PERFORMANCE OF HOT HGCDTE PHOTODIODES

Infrared photon detectors are typically operated at cryogenic temperatures to decrease the noise of the detector arising from various mechanisms associated with the narrow bandgap. The cooled technologies are expensive and, for many applications, are unattractive due to their prohibitive size, weight, and power signature. There have been considerable efforts to decrease system size, weight, and power consumption (SWaP) – in consequence, reducing the system's cost – to increase the operating temperature in high-operating-temperature (HOT) detectors. The ultimate goal is the fabrication of a detector with the dark current less than the system background flux current and with insignificant $1/f$ noise relative to the shot noise of the background flux. In 1999, the famous British scientist Tom Elliott and his co-workers wrote "that there is no fundamental obstacle to

TABLE 3.1  General Properties of Thermal Detectors

| Type | Temperature (K) | $D^*$ (cmHz$^{1/2}$/W) | NEP (WHz$^{1/2}$) | $\tau_{th}$ (ms) | Size (mm$^2$) |
|---|---|---|---|---|---|
| Silicon bolometer | 1.6 | | $3\times10^{-15}$ | 8 | 0.25–0.70 |
| Metal bolometer | 2–4 | $1\times10^8$ | | 10 | |
| Thermistor bolometer | 300 | $(1-6)\times10^8$ | | 1–8 | 0.01–10 |
| Germanium bolometer | 2–4 | | $5\times10^{-13}$ | 0.4 | 1.5 |
| Carbon bolometer | 2–4 | | $3\times10^{-12}$ | 10 | 20 |
| Superconducting bolometer (NbN) | 15 | | $2\times10^{-11}$ | 0.5 | $5\times0.25$ |
| Thermocouples | 300 | | $(2-10)\times10^{-10}$ | 10–40 | $0.1\times1$ to $0.3\times3$ |
| Thermopiles | 300 | $(1-3)\times10^8$ | | 3.3–10 | 1–100 |
| Pyroelectrics | 300 | $(2-5)\times10^8$ | | 10–100 | $2\times2$ |
| Golay cell | 300 | $1\times10^9$ | $6\times10^{-11}$ | 10–30 | 10 |

obtaining room temperature operation of photon detectors at room temperature with background-limited performance, even in reduced fields of view" [9]. In this section, we attempt to reconsider the performance of HOT photodetectors in the infrared spectral range.

In 2007, the Teledyne research group published an empirically derived equation, known as "Rule 07", for estimation of the dark current of P-on-n HgCdTe photodiodes *versus* the normalized wavelength-temperature product ($\lambda_c T$) [10]. This equation predicts the dark current density within a factor of 2.5 over a 13 order-of-magnitude range and is approximately (exact equation given in the reference [10]):

$$J_{dark} = 8367 \, exp\left( -\frac{1.44212q}{k\lambda_c T} \right) \text{ for } \lambda_c \geq 4.635 \text{ μm}, \tag{3.25}$$

and

$$J_{dark} = 8367 \, exp\left( -\frac{1.44212q}{k\lambda_c T} \left[ 1 - 0.2008 \left( \frac{4.635 - \lambda_c}{4.635\lambda_c} \right)^{0.544} \right] \right) \tag{3.26}$$

for $\lambda_c < 4.635$ μm,

where $\lambda_c$ is the cutoff wavelength in μm, $T$ is the operating temperature in K, $q$ is the electron charge, and $k$ is Boltzmann's constant (both of the latter being expressed in SI units). Rule 07 was developed for operating temperature cutoff wavelength products between 400 μmK and ~1700 μmK, and for operating temperatures above 77 K.

The Rule 07 metric is closely related to an Auger 1 diffusion-limited photodiode with n-type extrinsic doping concentration in the active region close to $10^{15}$ cm$^{-3}$. Any detector architecture that is limited by Auger 7 p-type diffusion, or by depletion currents, will not behave according to Rule 07. Rule 07 is also an excellent tool for a quick comparison of an $R_o A$ product of other material systems with HgCdTe. However, caution should be taken when expanding it to other parameters, such as detectivity and lower operating temperatures.

In the past decade, the Rule 07 metric has become very popular with the IR community for other technologies (especially to III-V barrier and type-II superlattice devices) as a reference level. However, at the present stage of technology, the fully depleted background-limited HgCdTe photodiodes can achieve the level of room-temperature dark current considerably lower than that predicted by Rule 07. The discussion below explains this statement exactly.

### 3.3.1 SRH Carrier Lifetime

The Shockley–Read–Hall (SRH) generation-recombination mechanism determines the carrier lifetimes in lightly doped n- and p-type HgCdTe, in which SRH centers are associated with residual impurities and native defects. From data gathered by Kinch et al. in 2005 [11], the measured values of carrier lifetimes for LWIR n-type HgCdTe range from 2 up to 20 μs at 77 K, regardless of doping concentration, for values below $10^{15}$ cm$^{-3}$. The values for mid-wave infrared (MWIR) material are typically slightly longer, in the range of 2–60 μs. However, several papers published in the past decade have shown the values of $\tau_{SRH}$ to be considerably larger in the low-temperature range and at low doping concentrations, above 200 μs up to even 50 ms, depending on the cutoff wavelength [12] (Table 3.2). The lowermost range of low doping that can be reproducibly generated in Teledyne growth HgCdTe epilayers by molecular beam epitaxy (MBE) is about $10^{13}$ cm$^{-3}$. In a recently published paper by Gravrand et al. [13], it has been shown that, for most tested MWIR photodiodes from LETI and Sofradir, the estimated SRH carrier lifetimes [from direct measurements (photoconductive or photoluminescence decay) as well as indirect estimations from current-voltage (I-V) characteristics], are in the range between 10 and 100 μs. These values are lower than previously estimated by US research groups; however, the latter were estimated for photodiodes with higher doping in the active region, above $10^{14}$ cm$^{-3}$. From recently published results [14], Teledyne has confirmed fabrication of depletion layer-limited P-i-N HgCdTe photodiodes, with SRH recombination centers having lifetimes in the range 0.5–10 ms.

All SRH lifetimes estimated for HgCdTe are usually carried out for temperatures below 300 K. Their extrapolation to 300 K, in order to predict the photodiode operation behavior, is questionable. In our estimation

TABLE 3.2  Summary of the SRH Carrier Lifetimes Determined on the Basis of I-V and FPA Characteristics (after Ref. [12])

|  | x composition | $\tau_{SRH}$ (μs) |
|---|---|---|
| LWIR | 0.225 | >100 at 60 K |
| MWIR | 0.30 | >1000 at 110 K |
| MWIR | 0.30 | ~50000 at 89 K |
| SWIR | 0.455 | >3000 at 180 K |

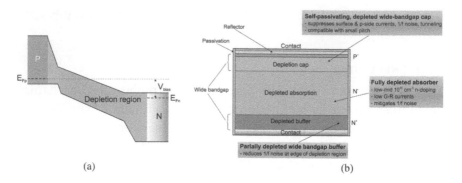

(a)                                              (b)

FIGURE 3.6    P-i-N photodiode: (a) energy band diagram under reverse bias, (b) heterojunction architecture.

we assume $\tau_{SRH}$ to equal 1 ms, which is supported by experimental data achieved by DRS and Teledyne research groups.

Figure 3.6(a) shows a schematic band diagram for a reverse-biased P-i-N heterostructure photodiode. The active region consists of an undoped i-region ($\nu$ region, low n-doping) sandwiched between a wider bandgap cap (P) and buffer (N) region [see Fig. 3.6(b)]. Very low doping in the absorber region (below $5 \times 10^{13}$ cm$^{-3}$) is required to allow full depletion at zero or a low value of reverse bias [15]. The surrounded wide-bandgap contact layers are designed to suppress the dark current generation from these regions and to suppress tunneling current under reverse bias. Moreover, fully depleted absorbers, surrounded by wide-bandgap regions, potentially reduce 1/$f$ and random telegraph noise. As previously mentioned, the fully depleted P-i-N structure is compatible with the small pixel size, achieving low crosstalk thanks to the built-in vertical electric field [12,15].

In P-i-N design, the choice of absorber thickness should be a trade-off between the response speed and quantum efficiency (or responsivity). To achieve short response times, the absorber thickness should be thin and fully depleted. For high quantum efficiency, the absorption region should be thick enough to effectively collect photogenerated carriers. However, to enhance quantum efficiency, while maintaining high response speed, an external resonant microcavity has been proposed. In this approach, the absorber is placed inside a cavity, so that a large portion of the photons can be absorbed, even with a small detection volume.

### 3.3.2 Dark Current Density

In general, for fully depleted P-i-N photodiodes, the limiting dark currents are diffusion currents in the N and P regions (depending on SRH

and Auger generations) and depletion current ruled only by SRH generation in the space charge region. The influence of radiative recombination
is still debatable but is not considered to be a limiting factor of small-pixel
HgCdTe photodiodes. Moreover, due to the photon recycling effect, the
influence of radiative recombination can be significantly reduced [16]. For
that reason, in our discussion, the radiative recombination is omitted.

The diffusion current of P-i-N HgCdTe photodiode structure arises
from the thermal generation of carriers in a thick, undepleted absorber
and is dependent on the Auger and SRH generation in n-type semiconductors [12]

$$J_{dif} = \frac{qn_i^2 t_{dif}}{n} \left( \frac{1}{\tau_{A1}} + \frac{1}{\tau_{SRH}} \right), \tag{3.27}$$

where $q$ stands for the electron charge, $n$ is the electron concentration, $t_{dif}$
is the diffusion region thickness, $n_i$ is the intrinsic carrier concentration,
$\tau_{A1}$ is the Auger 1 lifetime, and $\tau_{SRH}$ is the SRH lifetime. Auger 1 lifetime is
related to the hole, electron, and intrinsic carrier concentrations, and $\tau_{A1}$
is given by the equation:

$$\tau_{A1} = \frac{2\tau_{A1}^i n_i^2}{n(n+p)}, \tag{3.28}$$

where $p$ is the hole concentration and $\tau_{Ai}$ is the intrinsic Auger 1 lifetime.

For a low-temperature operation or a non-equilibrium active volume,
when the majority carrier concentration is held to be equal to the majority carrier doping level [and intrinsically generated majority carriers are
excluded ($p \ll n \approx N_{dop}$)], Eq. (3.28) becomes:

$$\tau_{A1} = \frac{2\tau_{A1}^i n_i^2}{n^2}. \tag{3.29}$$

The shortest SRH lifetime occurs through centers located approximately at the intrinsic energy level in the semiconductor bandgap. Then,
for the field-free region in an $n$ volume ($n \gg p$), $\tau_{SRH}$ is given by

$$\tau_{SRH} = \frac{\tau_{no} n_i + \tau_{po} (n + n_i)}{n}, \tag{3.30}$$

where $\tau_{no}$ and $\tau_{po}$ are the specific SRH lifetimes. At low temperatures,
where $n > n_i$, we have $\tau_{SRH} \approx \tau_{po}$. At high temperatures where $n \approx n_i$, we have

$\tau_{SRH} \approx \tau_{no} + \tau_{po}$. For a non-equilibrium active volume, $\tau_{SRH} \approx (\tau_{no} + \tau_{po})n_i/n$ exhibits a temperature dependence given by $n_i$.

A comparison of the Auger 1 and SRH lifetimes in equilibrium and non-equilibrium modes is shown in Fig. 3.7 for MWIR ($\lambda_c = 5$ μm) and LWIR ($\lambda_c = 10$ μm) HgCdTe with n-type doping concentration of $5 \times 10^{13}$ cm$^{-3}$. In calculations, the Hansen and Schmit analytical expression for the intrinsic carrier concentration was used [17]. The strong increase in SRH lifetimes for the non-equilibrium mode of operation at high temperatures is caused by the decreasing electron population of the SRH level, which, in consequence, results in the decrease of the minority carrier capture rates [12].

The second component is the depletion current arising from the portion of the absorber that becomes depleted. The depletion current density can be estimated by the following expression:

$$J_{dep} = \frac{qn_i t_{dep}}{\tau_{no} + \tau_{po}},$$  (3.31)

where $t_{dep}$ is the width of the depletion region.

The P-i-N HOT photodiode is characterized by useful properties at a reverse-biased operation. Figure 3.8 shows the calculated reverse-biased voltage which is required to completely deplete a 5-μm thick absorber doped at different doping levels. For the Rule 7 doping range of about $10^{15}$ cm$^{-3}$,

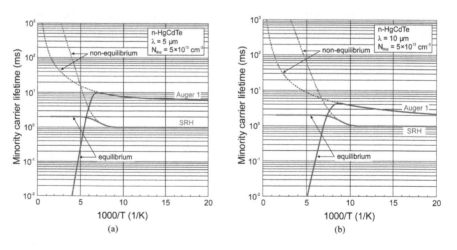

FIGURE 3.7 Equilibrium and non-equilibrium Auger 1 and SRH minority carrier lifetimes *versus* inverse temperature for MWIR and LWIR HgCdTe with n-type doping concentration of $5 \times 10^{13}$ cm$^{-3}$.

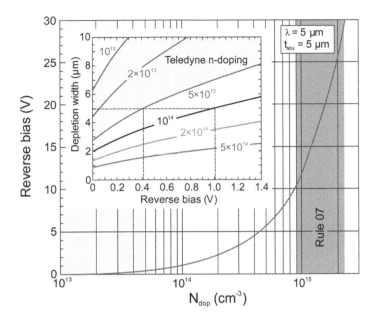

FIGURE 3.8 Calculated reverse-biased voltage *versus* doping concentration required to deplete a 5-μm thick MWIR HgCdTe absorber. Inset: Width of absorber depletion *versus* reverse-biased voltage and doping concentration.

a 5-μm thick absorber can be fully depleted by applying a relatively high reverse bias between 10 V and 30 V. On the other hand, for the range of doping reached presently at Teledyne (about $10^{13}$ cm$^{-3}$), the 5-μm thick absorber can be fully depleted for reverse bias from zero up to 0.4 V.

If P-i-N photodiode operates under reverse bias, the Auger suppression effect should be taken into account. This effect is important under HOT conditions, when $n_i \gg N_{dop}$. At non-equilibrium, large numbers of intrinsic carriers can be swept-out the absorber region. It is expected that this impact is larger for lower n-doping levels, since $n_i$ will be proportionately higher under these conditions. At very low levels of n-type doping (about $10^{13}$ cm$^{-3}$), the ultimate performance of P-i-N photodiode is limited by SRH recombination and neither Auger recombination nor Auger suppression.

As is shown in Fig. 3.9, for sufficiently long SRH-carrier lifetime in HgCdTe, the internal photodiode current is suppressed, and the performance is limited by the background radiation. The current density is shown at four background temperatures: 300, 200, 100, and 50 K. In [14], it is proposed to replace "Rule 07" with "Law 19". Law 19 corresponds exactly to the background-limited curve at room temperature. The internal

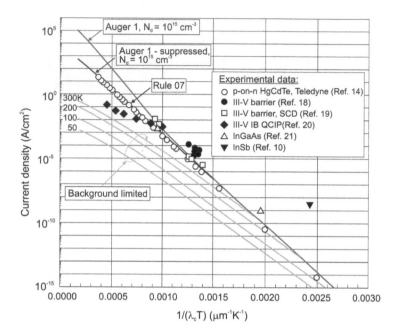

FIGURE 3.9  Current density of p-on-n HgCdTe photodiodes *versus* $1/(\lambda_c T)$ product (adapted after Ref. [13]). Experimental data are gathered from Teledyne and alternative technologies.

photodiode current can be several orders of magnitude below Rule 07 in dependence on a specific cutoff wavelength and operating temperature. It can also be seen that Rule 07 coincides well with a theoretically predicted curve for Auger-suppressed p-on-n photodiode with doping concentration in an active region equal to $N_d = 10^{15}$ cm$^{-3}$.

Figure 3.9 collates published experimental data for p-on-n HgCdTe photodiodes (Teledyne) [14] and for alternatives to HgCdTe material systems like III-V barrier detectors (Raytheon [18] and SCD [19]), operated at about 80 K, and room temperature inter-band quantum cascade infrared photodetectors IB QCIP [20]. It is easy to notice that experimental values for III-V barrier detectors are slightly poorer in comparison with those from p-on-n HgCdTe photodiodes, but III-V IB QCIPs operated at 300 K perform even better in the long wavelength spectral region. Figure 3.9 also shows representative data for both InSb ($\lambda_c = 5.3$ µm, $T = 78$ K) and InGaAs ($(\lambda_c = 1.7$ and $3.6$ µm, $T = 300$ K)) photodiodes. The InSb detector is characterized by several orders of magnitude higher dark current density than the HgCdTe one, although, for optimal InGaAs photodiodes, the dark current density is close to HgCdTe data [21].

Theoretical simulations presented in Fig. 3.9 show that the background limitation has the greatest impact on photodiode current density for small $1/(\lambda_c T)$ products, in other words, for photodiodes operated at the long wavelength region and under high operating-temperature conditions. HgCdTe photodiodes operated at low temperatures become generation-recombination-limited due to the influence of SRH centers having lifetimes in the millisecond range.

Figure 3.10 shows the current density calculated using Rule 07 (determined for diffusion limited P-on-n photodiodes) or Law 19 (which is exactly equal to the background radiation current density) as a function of temperature for the SWIR (3 μm), MWIR (5 μm), and LWIR (10 μm) absorbers.

If the fully depleted P-i-N photodiode is to be limited by the background radiation current, a certain minimal value of SRH lifetime is required. The calculations of SRH lifetime were made for which the depletion dark current is equal to the background radiation current:

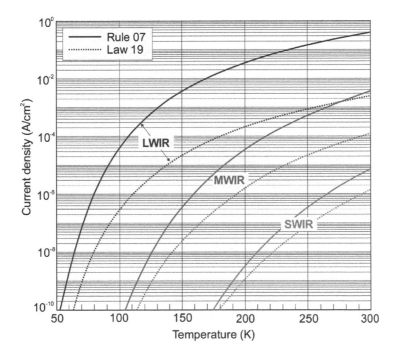

FIGURE 3.10   Calculated current density *versus* temperature relationship, using Law 19 and Rule 07 for SWIR (3 μm), MWIR (5 μm), and LWIR (10 μm) HgCdTe absorber.

$$J_{dep} = \frac{qn_i t_{dep}}{2\tau_{SRH}} = J_{BLIP}. \tag{3.32}$$

It was assumed that the 5-μm thick absorber is fully depleted.

The SRH lifetime, at which the fully depleted P-i-N photodiode reaches the BLIP limit, is presented in Fig. 3.11. As we can see, the SRH lifetime required to reach the BLIP limit decreases with increasing temperature, although fully depleted P-i-N photodiodes are particularly interesting under HOT conditions. Furthermore, for LWIR detectors, achieving BLIP performances is possible for shorter carrier lifetimes. At 300 K, these carrier lifetimes are 15 ms for SWIR (3 μm), 150 μs for MWIR (5 μm), and 28 μs for LWIR (10 μm) 5-μm thick fully depleted absorbers.

The Teledyne experimentally measured SRH lifetimes, extracted at 30 K for 10-μm cutoff HgCdTe, are reproducibly greater than 100 ms [15]. Despite the fact that the lifetimes at 300 K are likely to be a minimum of

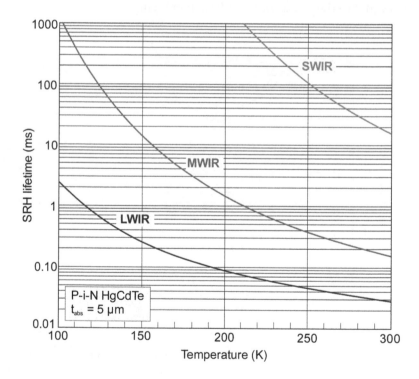

FIGURE 3.11 The SRH lifetime *versus* temperature relationship, for which the depletion dark current of the fully depleted P-i-N HgCdTe photodiode is equal to the background radiation current. The calculations are carried out for SWIR (3 μm), MWIR (5 μm), and LWIR (10 μm) absorbers.

10 times lower (due to the increase in thermal velocity, which increases the probability of carrier capture by a recombination center), there are still values of SRH lifetimes which enable it to reach the BLIP limit. This prediction is supported by theoretical simulation presented in [22].

### 3.3.3 Detectivity

The photodiode $D^*$ is specified on the basis of current responsivity, $R_i$, and noise current, $i_n$, and can be written as:

$$D^* = \frac{R_i}{i_n}. \tag{3.33}$$

For the non-equilibrium devices, the $i_n$ value can be calculated including thermal Johnson–Nyquist noise and shot noise, using the following expression

$$i_n = \sqrt{\frac{4kT}{R_d A} + 2qJ_{dark}}, \tag{3.34}$$

where $k$ is the Boltzmann constant, $R_d A$ is the dynamic resistance area product, and $J_{dark}$ is the dark current density.

The performance of P-i-N MWIR HgCdTe photodiode ($\lambda_c = 5$ μm) is presented in Fig. 3.12. Figure 3.12(a) shows the diffusion and depletion dark current components *versus* temperature relationship, assuming the value of the SRH carrier lifetime to be 1 ms, with an absorber thickness of 5 μm and doping of $5 \times 10^{13}$ cm$^{-3}$. For this doping level, a 5-μm thick absorber can be fully depleted at the reverse bias of 0.4 V (see Fig. 3.8). The diffusion component associated with the Auger 1 mechanism is eliminated because of the absence of majority carriers due to exclusion and extraction effects [23,24]. The background radiation calculated from $f$/3 optics has a decisive influence on the dark current. It should be mentioned here that the background flux current is defined by the total flux through the optics (limited by $f$/#), plus the flux from the cold shield. This effect is shown by increasing influence of the background-limited performance (BLIP) ($f$/3) on dark current at temperatures above 220 K.

As is shown in Fig. 3.12(a), the Teledyne Judson experimentally measured current densities, at the bias of −0.3 V, are close to the BLIP ($f$/3) curve, being located less than one order of magnitude above this limit.

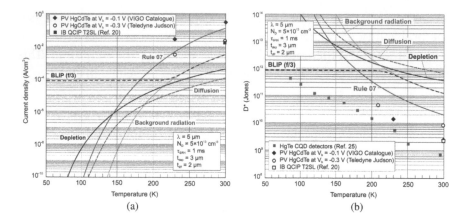

FIGURE 3.12 Performance of MWIR P-i-N HgCdTe photodiode with the value of $\tau_{SRH} = 1$ ms and absorber doping level of $5 \times 10^{13}$ cm$^{-3}$: (a) diffusion and depletion current components *versus* temperature, (b) detectivity *versus* temperature. The thickness of active region is $t = 5$ μm and consists of $t_{dif} = 2$ μm and $t_{dep} = 3$ μm. The experimental data are taken from different sources as marked. PV – photodiode; CQD – colloidal quantum dot (CQD); IB QCIP – inter-band quantum cascade infrared photodetector.

The current density at room temperature is even lower than that predicted by Rule 07. The measured current densities presented by VIGO are about one order of magnitude higher, although, in this case, they were measured at lower reverse bias, –0.1 V, with a less effective contribution from Auger suppression. It is interesting to note that the performance of IB QCIPs, based on type-II superlattices (T2SLs) fabricated with InAs/GaSb, coincide well with the upper experimental data for HgCdTe photodiodes at room temperature [20].

Figure 3.12(b) shows calculated detectivity *versus* temperature relationship for a MWIR P-i-N HgCdTe photodiode, assuming identical parameters taken in calculation, as presented in Fig. 3.12(a) ($\lambda_c = 5$ μm, $\tau_{SRH} = 1$ ms, $t = 5$ μm, $N_{dop} = 5 \times 10^{13}$ cm$^{-3}$). The current responsivity was calculated, assuming quantum efficiency (QE) = 1 (although typical QE reaches a reasonable value of about 0.7). As is shown, for a MWIR photodiode with a 5-μm cutoff wavelength and low doping in the active region, $D^*$ is limited by background and is about one order of magnitude higher than predicted by Rule 07. The experimental data given for HgCdTe photodiodes in Teledyne Judson and VIGO catalogues are more than one order of magnitude below the background flux limitation for the $f/3$ optics.

## 3.4 INTERBAND QUANTUM CASCADE INFRARED PHOTODETECTORS (IB QCIPS)

The performance of conventional p-n junction LWIR HgCdTe HOT photodiodes, with active doping concentrations above $10^{16}$ cm$^{-3}$, is poor, due to a low quantum efficiency (as a result of low diffusion length and weak absorption of radiation) and a low dynamic resistance. Only charge carriers, that are photogenerated at distances shorter than the diffusion length from junction, can be collected. The absorption depth of long wavelength IR radiation ($\lambda > 5$ µm) is longer than the diffusion length and therefore, only a limited fraction of the photogenerated charge contributes to the quantum efficiency. For example, estimates show that the ambipolar diffusion length in a 10.6-µm photodiode is less than 2 µm while the absorption depth is $\approx$12 µm. In consequence, the quantum efficiency reduces to $\approx$15% for a single pass of radiation through the detector [2].

To overcome the above problems, multiple heterojunction photovoltaic devices, with short elements connected in series, have been proposed. An example is a device, introduced in 1995, with the junctions' planes perpendicular to the substrate, as shown in Fig. 3.13(a). The multiple heterojunction device consists of a structure based on backside-illuminated n$^+$-p-P photodiodes. Such a device was characterized by high voltage responsivity and fast response time, but it suffered from a nonuniform response across the active area and dependence of the response on polarization of incident radiation.

A more promising option appears to be the stacked tunnel junctions, connected in series, shown in Fig. 3.13(b). This idea is similar to multijunction solar cells. Potentially, this device is capable of achieving both high quantum efficiency, large differential resistance, and fast response. As is shown, each cell consists of a p-type doped absorber and heavily doped N$^+$ and P$^+$ heterojunction contacts. The incoming radiation is

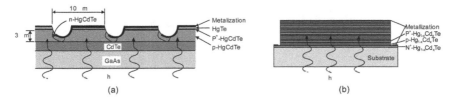

FIGURE 3.13    Backside-illuminated multiple HgCdTe heterojunction devices: (a) junctions' planes perpendicular to the surface, and (b) 4-cell-stacked multiple detector (after Ref. [26]).

absorbed only in absorber regions, whereas the heterojunction contacts collect the photogenerated charge carriers. However, a practical problem is the resistance of the adjacent $N^+$ and $P^+$ regions.

In the past decade, new types of multi-stage IR detectors, based on III-V semiconductors, have been proposed, which are now divided into two classes: (i) inter-sub-band (IS) unipolar quantum cascade infrared photodetectors QCIPs [27], and (ii) inter-band (IB) ambipolar QCIPs [28]. The earliest investigations of IS QCIPs, which evolved from the quantum cascade lasers (QCLs), began about two decades ago. However, in comparison with forward-biased QCLs, where one injected electron creates many photons, the QCIPs operating under zero-bias conditions are based on the creation of one electron by photons, meaning that the photosignal is determined by an individual absorber in one stage (absorber + relaxation + tunneling layers; Fig. 3.14). Each individual absorber is sandwiched between the electron bandgaps which are substantially wider than the bandgap of the absorber, forming a cascade stage. Each of the individual absorbers in a QCIP is designed to be shorter than the diffusion length, to efficiently collect all photogenerated charge carriers.

The total photocurrent is independent of the number of detector's constituent stages, and the photons absorbed in the following stages do not enlarge the photocurrent but only maintain the current continuity flowing through the device. Discrete absorber design ensures that the photogenerated electrons recombine with holes in the next stage within a short transport distance, circumventing the diffusion length limitation in traditional detectors with thick absorbers. In order to suppress noise, multiple discrete short absorbers are connected in series, so that the total thickness of the absorbers can be even higher than the absorption depth. The noises are suppressed by using shorter individual absorbers and a larger number

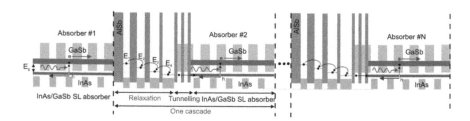

FIGURE 3.14 Schematic illustration of an IB QCIP with T2SL InAs/GaSb absorbers, GaSb/AlSb tunneling and InAs/AlSb relaxation layers (after Ref. [28]).

of cascade stages. The detectivity of QCIP, limited by Johnson noise and electrical shot noise, can be described [27] by:

$$D^* = \frac{\eta \lambda q}{hc} \left( \frac{4kT}{NR_0 A} + \frac{2qI_{dark}}{N} \right)^{-1/2}, \qquad (3.35)$$

where $R_0 A$ is the dynamic resistance at a zero bias, corresponding to one period of QCIP, and $N$ is the number of periods. Equation (3.35) shows that the $S/N$ ratio is proportional to $\sqrt{N}$.

It appears that the relatively much longer carrier lifetime in IB QCIPs (nanosecond range) in comparison with IS QCIPs (picosecond range) determines the higher performance of the IB devices. In consequence, the saturation current density of IB QCPs is almost two orders of magnitude lower than that achieved by IS QCPs [20].

The MBE growth of IB QCIPs, with InAs/GaSb T2SL active regions, is challenging, involving many interfaces and strained thin layers in their structures. Nevertheless, significant progress has been made for T2SL detectors, particularly in the LWIR spectrum and at high temperatures. They combine the advantages of inter-band optical transitions with the excellent carrier transport properties of the inter-band cascade laser structures.

IB QCIPs are divided into two configuration types: current-matched, where the photocurrent is designed to be equal for all cascade stages, and non-current-matched [29]. Hinkey and Yang reported that multiple-stage architecture, with equal absorber lengths in each stage, offers the potential for significant improvement in detectivity when $\alpha L \leq 0.2$, where $\alpha$ is the absorption coefficient and $L$ is the diffusion length [30]. The optimal number of stages is directly related to the single absorber thickness, $d$, and the absorption coefficient, and the first-order approximation could be expressed as $N = (2\alpha d)^{-1}$. Compared to the current-matched IB QCIPs, the non-current-matched IB QCIPs, with identical absorbers in all stages, are simpler to design and implement. The drawback of non-current-matched architecture is the reduced responsivity as a result of significant light attenuation. However, high electrical gain, recently observed in these structures under high-temperature operation, could at least partially compensate for the reduction of the responsivity due to light attenuation [31,32].

At present, HgCdTe is the most widely used variable gap semiconductor for IR photodetectors, including uncooled operations, and stands as a reference for alternative technologies. Figure 3.15 (dashed lines) demonstrates

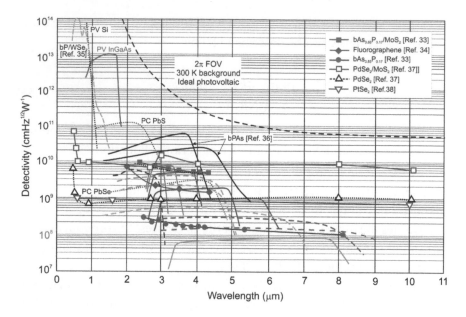

FIGURE 3.15  Room-temperature spectral detectivity curves of commercially available photodetectors [PV Si and InGaAs, PC PbS and PbSe, HgCdTe photodiodes (solid lines)]. The spectral detectivity curves of newly emerging T2SL IB QCIPs are marked by dashed lines [28]. In addition, the experimental data for different types of 2D material photodetectors are included [33–38]. PC – photoconductor, PV – photodiode.

that bipolar devices, based on type-II InAs/GaSb IB superlattice absorbers, are a good candidate for detectors operating at near-room temperature. The estimated Johnson-noise limited detectivities under zero bias for IB cascade photodetectors with T2SL InAs/GaSb absorbers (based on the measured $R_0A$ product and responsivity) are comparable with those available for commercial HgCdTe photovoltaic detectors. We can see that the performance of both types of detectors is comparable in short wavelength infrared regions, but inter-band cascade detectors outperform commercially available uncooled HgCdTe detectors with a similar LW cutoff wavelength. In addition, due to strong covalent bonding of III-V semiconductors, IB QCIPs can be operated at temperatures up to 400°C, which is not feasible for their HgCdTe counterparts.

Figure 3.16 collates the highest detectivity values published in the literature for different types of single-element photodetectors operated at room temperature. This fact should be clearly emphasized since detectivity data, marked for commercial photodetectors, are typical for pixels of infrared

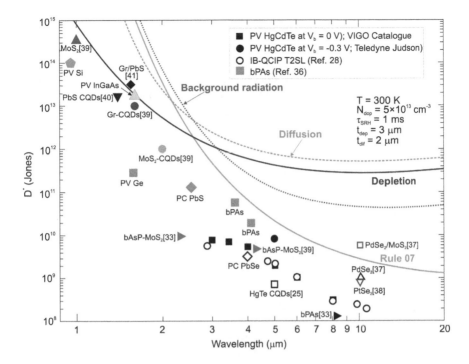

FIGURE 3.16 Dependence of detectivity on the wavelength for the commercially available room-temperature-operating infrared photodetectors (PV Si and Ge, PV InGaAs, PC PbS and PbSe, PV HgCdTe). Experimental data are also included for IB QCIP T2SLs, different type of 2D material and colloidal quantum dot (CQD) photodetectors, taken from the literature as marked in the square brackets. The theoretical curves are calculated for P-i-N HOT HgCdTe photodiodes, assuming the value of $\tau_{SRH} = 1$ ms, the absorber doping level of $5 \times 10^{13}$ cm$^{-3}$ and the thickness of the active region, $t = 5$ μm. PC – photoconductor; PV – photodiode.

focal plane arrays. Figure 3.16 also identifies the fundamental indicator for future trends in the development of HOT infrared photodetectors. At the present stage of HgCdTe technology, the semiempirical rule Rule 07 is found not to fulfill primary expectations. It is shown that the detectivity of low-doping P-i-N HgCdTe ($5 \times 10^{13}$ cm$^{-3}$) photodiodes, operating at room temperature in a spectral band greater than 3 μm, is limited by background radiation (with $D^*$ level above $10^{10}$ Jones, not limited by the detector itself) but can be improved by more than one order of magnitude in comparison with that predicted by Rule 07. Between different material systems used in the fabrication of HOT LWIR photodetectors, only the HgCdTe ternary alloy can fulfill the required expectations, namely low doping concentration ($10^{13}$ cm$^{-3}$) and high SRH carrier lifetime (above 1 ms). In this context, it will be rather difficult to rival 2D material photodetectors and CQD photodetectors with HgCdTe photodiodes. The above estimations provide further encouragement for achieving low-cost and high-performance MWIR and LWIR HgCdTe focal plane arrays operated under HOT conditions. The performance of T2SL IB QCIPs is close to that of HgCdTe photodiodes, and quantum cascade photodetectors can be operated in temperatures above 300 K; however, their disadvantage is the challenging technology involved and the higher cost of fabrication.

## REFERENCES

1. J. Piotrowski and A. Rogalski, Comment on "Temperature limits on infrared detectivities of InAs/In$_x$Ga$_{1-x}$Sb superlattices and bulk Hg$_{1-x}$Cd$_x$Te" [*Journal of Applied Physics* **74**, 4774 (1993)], *Journal of Applied Physics* **80**(4), 2542–2544 (1996).
2. J. Piotrowski and A. Rogalski, *High-Operating-Temperature Infrared Photodetectors*, SPIE Press, Bellingham, 2007.
3. A. Rogalski, "Quantum well photoconductors in infrared detector technology", *Journal of Applied Physics* **93**, 4355 (2003).
4. A. Rogalski, M. Kopytko, and P. Martyniuk, *Antimonide-based Infrared Detectors: A New Perspective*, SPIE Press, Bellingham, 2018.
5. A. Smith, F.E. Jones, and R.P. Chasmar, *The Detection and Measurement of Infrared Radiation*, Clarendon, Oxford, 1968.
6. W. Kruse, L.D. McGlauchlin, and R.B. McQuistan, *Elements of Infrared Technology*, Wiley, New York, 1962.
7. J.D. Vincent, S.E. Hodges, J. Vampola, M. Stegall, and G. Pierce, *Fundamentals of Infrared and Visible Detector Operation and Testing*, Wiley, Hoboken, 2016.
8. P.G. Datskos, "Detectors—Figures of merit", in *Encyclopedia of Optical Engineering*, pp. 349–357, ed. R. Driggers, Marcel Dekker, New York, 2003.
9. C.T. Elliott, N.T. Gordon, and A.M. White, "Towards background-limited, room-temperature, infrared photon detectors in the 3–13 μm wavelength range", *Applied Physics Letters* **74**(9), 2881–2883 (1999).
10. W.E. Tennant, D. Lee, M. Zandian, E. Piquette, and M. Carmody, "MBE HgCdTe technology: A very general solution to IR detection, described by 'Rule 07', a very convenient heuristic", *Journal of Electronic Materials* **37**, 1406–1410 (2008).
11. M.S. Kinch, F. Aqariden, D. Chandra, P.-K. Liao, H.F. Schaake, and H.D. Shih, "Minority carrier lifetime in p-HgCdTe", *Journal of Electronic Materials* **34**, 880–884 (2005).
12. M.A. Kinch, *State-of-the-Art Infrared Detector Technology*, SPIE Press, Bellingham, 2014.

13. O. Gravrand, J. Rothman, B. Delacourt, F. Boulard, C. Lobre, P.H. Ballet, J.L. Santailler, C. Cervera, D. Brellier, N. Pere-Laperne, V. Destefanis, and A. Kerlain, "Shockley-Read-Hall lifetime study and implication in HgCdTe photodiodes for IR detection", *Journal of Electronic Materials* **47**(10), 5680–5690 (2018).

14. D. Lee, P. Dreiske, J. Ellsworth, R. Cottier, A. Chen, S. Tallarico, A. Yulius, M. Carmody, E. Piquette, M. Zandian, and S. Douglas, "Law 19—The ultimate photodiode performance metric", *Extended Abstracts. The 2019 U.S. Workshop on the Physics and Chemistry of II-VI Materials*, pp. 13–15, 2019.

15. D. Lee, M. Carmody, E. Piquette, P. Dreiske, A. Chen, A. Yulius, D. Edwall, S. Bhargava, M. Zandian, and W.E. Tennant, "High-operating temperature HgCdTe: A vision for the near future", *Journal of Electronic Materials* **45**(9), 4587–4595 (2016).

16. M. Kopytko, K. Jóźwikowski, P. Martyniuk, and A. Rogalski, "Photon recycling effect in small pixel p-i-n HgCdTe long wavelength infrared photodiodes", *Infrared Physics and Technology* **97**, 38–42 (2019).

17. G.L. Hansen and J.L. Schmit, "Calculation of intrinsic carrier concentration in $Hg_{1-x}Cd_xTe$", *Journal Applied Physics* **54**, 1639–1640 (1983).

18. D.R. Rhiger, "Performance comparison of long-wavelength infrared type II superlattice devices with HgCdTe", *Journal of Electronic Materials* **40**, 1815–1822 (2011).

19. P.C. Klipstein, E. Avnon, D. Azulai, Y. Benny, R. Fraenkel, A. Glozman, E. Hojman, O. Klin, L. Krasovitsky, L. Langof, I. Lukomsky, M. Nitzani, I. Shtrichman, N. Rappaport, N. Snapi, E. Weiss, and A. Tuito, "Type II superlattice technology for LWIR detectors", *Proceedings of SPIE* **9819**, 98190T (2016).

20. W. Huang, S.M.S. Rassel, L. Li, J.A. Massengale, R.Q. Yang, T.D. Mishima, and M.B. Santos, "A unified figure of merit for interband and intersubband cascade devices", *Infrared Physics and Technology* **96**, 298–301 (2019).

21. A. Rogalski and R. Ciupa, "Performance limitation of short wavelength infrared InGaAs and HgCdTe photodiodes", *Journal of Electronic Materials* **28**(6), 630–636 (1999).

22. A. Rogalski, M. Kopytko, P. Martyniuk, "Performance prediction of p-i-n HgCdTe long-wavelength infrared HOT photodiodes", *Applied Optics* **57**(18), D11–D19 (2018).

23. T. Ashley and C.T. Elliott, "Non-equilibrium mode of operation for infrared detection", *Electronics Letters* **21**(10), 451–452 (1985).

24. C.T. Elliott, "Non-equilibrium mode of operation of narrow-gap semiconductor devices", *Semiconductor Science and Technology* **5**, S30–7 (1990).

25. P. Guyot-Sionnest, M.M. Ackerman, and X. Tang, "Colloidal quantum dots for infrared detection beyond silicon", *The Journal of Chemical Physics* **151**, 060901-18 (2019).

26. J. Piotrowski and A. Rogalski, "Uncooled long wavelength infrared photon detectors", *Infrared Physics and Technology* **46**, 115–131 (2004).

27. A. Gomez, M. Carras, A. Nedelcu, E. Costard, X. Marcadet, V. Berger, "Advantages of quantum cascade detectors", *Proceedings of SPIE* **6900**, 69000J-1–14 (2008).
28. A. Rogalski, P. Martyniuk, and M. Kopytko, "Type-II superlattice photodetectors versus HgCdTe photodiodes", *Progress in Quantum Electronics* **68**, 100228 (2019).
29. W. Huang, L. Li, L. Lei, J.A. Massengale, R.Q. Yang, T.D. Mishima, and M.B. Santos, "Electrical gain in interband cascade infrared photodetectors", *Journal of Applied Physics* **123**, 113104 (2018).
30. R.T. Hinkey and R.Q. Yang, "Theory of multiple-stage interband photovoltaic devices and ultimate performance limit comparison of multiple-stage and single-stage interband infrared detectors", *Journal of Applied Physics.* **114**, 104506-118 (2013).
31. W. Huang, L. Lei, L. Li, J.A. Massengale, R.Q. Yang, T.D. Mishima, and M.B. Santos, "Current-matching versus non-current-matching in long wavelength interband cascade infrared photodetectors", *Journal of Applied Physics* **122**, 083102 (2017).
32. L. Lei, L. Li, H. Lotfi, H. Ye, R.Q. Yang, T.D. Mishima, M.B. Santos, and M.B. Johnson, "Mid wavelength interband cascade infrared photodetectors with superlattice absorbers and gain", *Optical Engineering* **57**(1), 011006 (2018).
33. M. Long, A. Gao, P. Wang, H. Xia, C. Ott, C. Pan, Y. Fu, E. Liu, X. Chen, W. Lu, T. Nilges, J. Xu, X. Wang, W. Hu, F. Miao, "Room temperature high-detectivity mid-infrared photodetectors based on black arsenic phosphorus", *Science Advances* **3**, e1700589 (2017).
34. S. Du, W. Lu, A. Ali, P. Zhao, K. Shehzad, H. Guo, L. Ma, X. Liu, X. Pi, P. Wang, H. Fang, Z. Xu, C. Gao, Y. Dan, P. Tan, H. Wang, C.-T. Lin, J. Yang, S. Dong, Z. Cheng, E. Li, W. Yin, J. Luo, B. Yu, T. Hasan, Y. Xu, W. Hu, and X. Duan, "A broadband fluorographene photodetector", *Advanced Materials* **29**, 1700463 (2017).
35. L. Ye, P. Wang, W. Luo, F. Gong, L. Liao, T. Liu, L. Tong, J. Zang, J Xu, and W. Hu, "Highly polarization sensitive infrared photodetector based on black phosphorus-on-WSe$_2$ photogate vertical heterostructure", *Nano Energy* 37, 53–60 (2017).
36. M. Amani, E. Regan, J. Bullock, G.H. Ahn, and A. Javey, "Mid-wave infrared photoconductors based on black phosphorus-arsenic alloys", *ACS Nano* **11**, 11724–11731 (2017).
37. M. Long, Y. Wang, P. Wang, X. Zhou, H. Xia, C. Luo, S. Huang, G. Zhang, H. Yan, Z. Fan, X. Wu, X. Chen, W. Lu, and W. Hu, "Palladium diselenide long-wavelength infrared photodetector with high sensitivity and stability", *ACS Nano* **13**, 2511–2519 (2019).
38. X. Yu, P. Yu, D. Wu, B. Singh, Q. Zeng, H. Lin, Wu Zhou, J.Lin, K. Suenaga, Z. Liu, Q.J. Wang, "Atomically thin noble metal dichalcogenide: a broadband mid-infrared semiconductor", *Nature Communications* **9**, 1545 (2018).

39. G. Konstantatos, "Current status and technological prospect of photodetectors based on two-dimensional materials", *Nature Communications* **9**, 5266 (2018).
40. G. Konstantatos and E.H. Sargent, "Solution-processed quantum dot photodetectors", *Proceedings of IEEE* **97**(10), 1666–1683 (2009).
41. G. Konstantatos, M. Badioli, L. Gaudreau, J. Osmond, M. Bernechea, F.P. Garcia de Arquer, F. Gatti, and F.H.L. Koppens "Hybrid graphene-quantum dot phototransistors with ultrahigh gain", *Nature Nanotechnology* **7**, 363–368 (2012).

# CHAPTER 4

# Focal Plane Arrays

T HE BASIC CONCEPT OF a modern thermal-imaging system is to form a real image of the infrared (IR) scene, detect any variation in the imaged radiation, and, by suitable electronic processing, create a visible representation of this variation analogous to that achieved by conventional television cameras. IR camera construction is similar to that of a digital video camera. Detectors are only a part of usable sensor systems. Instead of charge-coupled device (CCD)/complementary metal-oxide semiconductor (CMOS) image arrays that video and digital still cameras use, the IR camera detector (Fig. 4.1) is a focal plane array (FPA) of micrometer-sized pixels made of various material sensitive to IR wavelengths. Once a detector is selected, optics (lens) material and filters can be selected to somewhat alter the overall response characteristics of an IR camera system.

The term "focal plane array" (FPA) refers to an assemblage of individual detector picture elements ("pixels") located at the focal plane of an imaging system. Although the definition could include 1D ("linear") arrays as well as 2D arrays, it is most frequently applied to the latter. Usually, the optics part of an optoelectronic image device is restricted to focusing of the image onto the detector's array. These so-called "staring arrays" are scanned electronically, usually by circuits integrated with the arrays. The architecture of detector-readout assemblies has assumed a number of forms which have been described in detail [1–4]. The types of readout integrated circuits (ROICs) include the function of pixel deselecting, antiblooming on each pixel, subframe imaging, and output preamplifiers, and may include yet other functions.

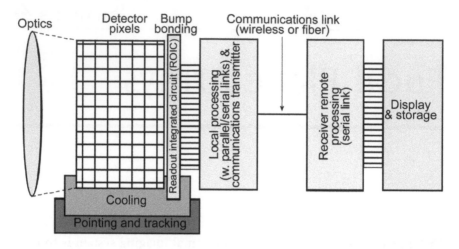

FIGURE 4.1 Schematic representation of an imaging system, showing important sub-systems.

FPA technology has revolutionized many kinds of imaging, from $\gamma$ rays to the terahertz and even radio waves; the rate at which images can be acquired has increased by more than a factor of a billion in many cases. Figure 4.2 illustrates the trend in array size over the past 50 years. Imaging IR FPAs have been developing in-line with the ability of silicon integrated circuit (IC) technology to read and process the array signals, and also to display the resulting image. The progress in IR arrays has also been steady, mirroring the development of dense electronic structures, such as dynamic random-access memories (DRAMs). FPAs have had nominally the same development rate as DRAM ICs, which have followed Moore's Law with a doubling-rate period of approximately 18 months, although with FPAs lagging DRAMs by about 5–10 years. The 18-month doubling time is evident from the slope of the graph presented in the Fig. 4.3, which shows the log of the number of pixels per array as a function of the first year of commercial availability. Array size will continue to increase but perhaps at a rate that falls below the Moore's Law curve [6]. An increase in array size is already technically feasible [7]. However, the market forces that have demanded larger arrays are not as strong now that the megapixel barrier has been broken. CCDs greater than three gigapixels offer the largest formats (Fig. 4.2).

Astronomers were the driving force towards the day when the opto-electronic arrays could match the size of photographic film. Since large arrays dramatically multiply the data output of a telescope system, the

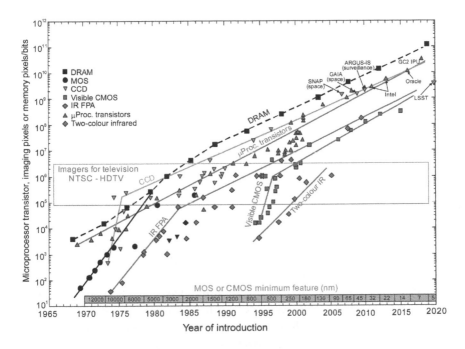

FIGURE 4.2 Imaging array formats compared with the complexity of silicon microprocessor technology and dynamic random-access memory (DRAM), as indicated by transistor count and memory bit capacity (adapted after Ref. [5] with completions). The timeline design rule of MOS/CMOS features is shown at the bottom. Note the rapid rise of CMOS imagers which are challenging CCDs in the visible spectrum. Infrared arrays with size greater than 100 megapixels are now available for astronomy applications. Imaging formats of many detector types have progressed beyond that required for high-definition TV.

development of large-format mosaic sensors of high sensitivity for ground-based astronomy is the goal of many astronomic observatories around the world. Raytheon manufactured a 4×4 mosaic of 2 k×2 k HgCdTe sensor chip assemblies (SCAs) with 67 million pixels and assisted in assembling it to the final focal-plane configuration to survey the entire sky in the southern hemisphere at four IR wavelengths [6]. Also, Teledyne fabricated a large detector system for the Euclid Near-Infrared Spectrometer and Photometer (NISP) instrument which consists of a 4×4 mosaic focal plane of 16 H2RG (2 k×2 k pixels) SCAs and 16 SIDECAR ASIC Sensor Chip Electronics (SCE) modules. These SCAs were made with substrate-removed molecular beam epitaxy (MBE) HgCdTe material system with a 2.3-μm cutoff wavelength. The SCAs are mounted on a buttable molybdenum package that enables close-packing of the 16 flight SCAs in the NISP

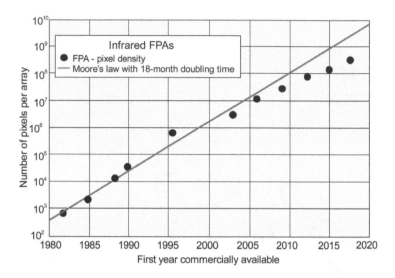

FIGURE 4.3   The number of pixels on an infrared array has been growing exponentially, in accordance with Moore's Law, for 40 years with a doubling time of approximately 18 months. The pixel density of infrared arrays will keep increasing with time but perhaps slower than that predicted by Moore's curve.

FIGURE 4.4   The top and bottom sides of the Euclid H2RG SCA molybdenum flight package (after Ref. [8]).

focal plane (Fig. 4.4). The mass of a fully assembled H2RG SCA is less than 115 grams [8].

The trend of increasing pixel number is likely to continue in the area of large-format arrays. It is predicted that a focal plane of 100 megapixels and larger will be possible, constrained by only budgets but not by technology

[9]. This increasing trend will be continued using a close-butted mosaic of several SCAs. "Butting" refers to tiling closely together separate pieces of semiconductor to produce one large sensitive array operated as a single image sensor. In most cases, butting is used to make imagers that are larger than the largest imager a single wafer can hold.

A number of architectures are used in the development of IR FPAs [10–14]. In general, they may be classified as hybrid or monolithic, but these distinctions are often not as important as proponents or critics state them to be. The central design questions involve performance advantages *versus* ultimate producibility. Each application may favor a different approach, depending on the technical requirements, projected costs, and schedule.

## 4.1 MONOLITHIC ARRAYS

In the monolithic approach, both detection of light and signal readout (multiplexing) is done in the detector material, rather than in an external readout circuit. The integration of detector and readout onto a single monolithic piece reduces the number of processing steps, increases yields, and reduces costs. Common examples of these FPAs in the visible and near-infrared (0.7–1.0 μm) regions are found in camcorders and digital cameras. Two generic types of silicon technology provide the bulk of devices in these markets: charge-coupled devices (CCDs) and complementary metal-oxide semiconductor (CMOS) imagers. CCD technology has achieved the highest pixel counts or largest formats with numbers greater than $10^9$ (Fig. 4.2). This approach to image acquisition was first proposed in 1970 in a paper written by Bell Lab researchers W.S. Boyle and G.E. Smith [15]. CMOS imagers are also rapidly moving to large formats and, at present, are competing with CCDs for the large-format applications. Figure 4.5 shows different architectures of monolithic FPAs.

### 4.1.1 CCD Devices

The basic element of a monolithic CCD array is a metal-insulator-semiconductor (MIS) structure. Used as part of a charge-transfer device, an MIS capacitor detects and integrates the generated photocurrent. Although most imaging applications tend to require high charge-handling capabilities in the unit cells, an MIS capacitor, fabricated from a narrow-bandgap semiconductor material (e.g., HgCdTe and InSb), has a limited charge capacity because of its low background potential, as well as more severe problems involving noise, tunneling effects, and charge trapping when shifting charge through the narrow-bandgap CCD to accomplish the readout function.

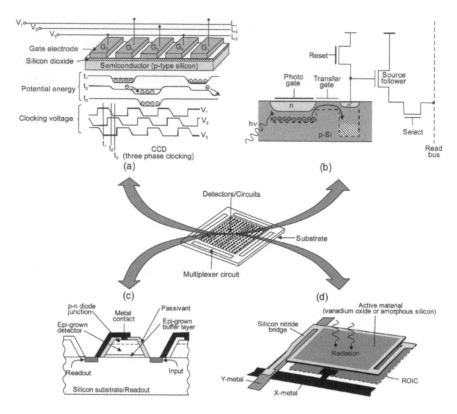

FIGURE 4.5  Monolithic focal plane arrays: (a) CCD, (b) CMOS, (c) heteroepitaxy-on-silicon, and (d) microbolometer.

Because of the non-equilibrium operation of the MIS detector, much larger electric fields are set up in the depletion region than in the p-n junction, resulting in a defect-related tunneling current that is orders of magnitude larger than the fundamental dark current. The MIS detector required much higher-quality material than was used in p-n junction detectors, something which has still not been achieved. So, although efforts have been made to develop monolithic FPAs, using narrow-bandgap semiconductors, silicon-based FPA technology is the only mature technology with respect to fabrication yield and attainment of near-theoretical sensitivity. Examples of fully monolithic FPAs with full TV resolution have been the commercially available PtSi Schottky-barrier FPAs. However, the performance of monolithic PtSi Schottky-barrier FPAs reached a plateau about 10 years ago and their further development was stopped in the late 2010s.

The CCD technology is very mature with respect to fabrication yield and attainment of near-theoretical sensitivity. The CCD technique relies on the optoelectronic properties of a well-established semiconductor architecture, namely the metal-oxide-semiconductor (MOS) capacitor. A MOS capacitor typically consists of an extrinsic silicon substrate, on which is grown an insulating layer of silicon dioxide ($SiO_2$). When a bias voltage is applied across a p-type MOS structure, majority charge carriers (holes) are pushed away from the $Si-SiO_2$ interface directly below the gate, leaving a region depleted of positive charge and available as a potential energy well for any mobile minority charge carriers (electrons) (Fig. 4.5a). Electrons generated in the silicon through absorption (charge generation) will collect in the potential-energy well under the gate (charge collection). Linear or two-dimensional arrays of these MOS capacitors can therefore store images in the form of trapped charge carriers beneath the gates. The accumulated charges are transferred from the potential well to the next well, by using sequentially shifted voltage on each gate (charge transfer). One of the most successful voltage-shifting schemes is called three-phase clocking. Column gates are connected to the separate voltage lines ($L_1$, $L_2$, $L_3$) in contiguous groups of three ($G_1$, $G_2$, $G_3$). The setup enables each gate voltage to be separately controlled.

Figure 4.6(a) shows the schematic circuit for a typical CCD imager. The photogenerated carriers are first integrated into an electronic well at the pixel and are subsequently transferred to slow and fast CCD shift registers. At the end of the CCD register, charge-carrying information on the received signal can be read out and converted into a useful signal (charge measurement).

The process of readout from the CCD consists of two parts:

- Moving charge packets (representing pixel values) around the sensor, and

- Converting the charge packet values into output voltages.

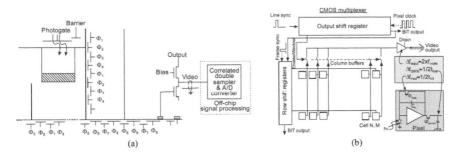

FIGURE 4.6 Typical readout architecture of (a) CCD and (b) CMOS images.

The charge-to-voltage converter at the CCD output is basically a capacitor with a single- or multistage voltage follower and a switch to preset the capacitor voltage to a "known" level. In simplest video systems, the switch is closed at the beginning of each pixel readout, which presets the capacitor voltage as well as the output level. After the pixel charge packet is transferred to the capacitor, its voltage changes and the output signal represents the pixel value. Due to the switch's finite residual conductivity, the capacitor is pre-charged to an unknown value and it adds the output signal. A way to compensate for this pre-charge uncertainty is the readout technique method, namely correlated double sampling (CDS). In this method, the output signal is sampled twice for each pixel, just after the pre-charging capacitor and after the pixel charge packet is added.

Figure 4.7 shows a preamplifier, in this example the source follower per detector (SFD), the output of which is connected to a clamp circuit. The output signal is initially sampled across the clamp capacitor during the onset of photon integration (after the detector is reset). The action of the clamp switch and the capacitor subtracts any initial offset voltage from the output waveform. Because the initial sample is made before significant photon charge has been integrated, by charging the capacitor, the final integrated photon signal swing is unaltered. However, any offset voltage or drift present at the beginning of integration is, by the action of the circuit, subtracted from the final value. This process of sampling each pixel twice, once at the beginning of the frame and again at the end, and providing the difference, is called CDS. More information about readout techniques used in CCD devices (CDS, floating diffusion amplifier in each pixel, and floating gate amplifier) are described in detail elsewhere, e.g., in references [16–21].

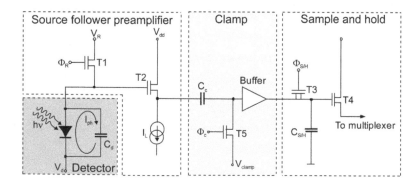

FIGURE 4.7    Correlated double sampling circuit.

The first CCD imager sensors were developed over 40 years ago, primarily for television analog image acquisition, transmission, and display. With increasing demand for digital image data, the traditional analog raster scan output of image sensors is of limited use, and there is a strong motivation to fully integrate the control, digital interface, and image sensor on a single chip.

The most popular CCD consists of a silicon sensor operating in visible and near-infrared (NIR) wavelength ranges. These spectra could be extended into the UV, using delta doping and antireflection coating. In this way, device stability and external quantum efficiency of up to 50–90% at wavelengths of 200–300 nm are obtained. CCD for scientific applications are routinely made with pixel counts exceeding 20 megapixels, and visible 50-megapixel arrays are now available with digital output shaving ROIC noise levels of less than 10 electrons and offering a sensitivity advantage over consumer products [22].

## 4.1.2 CMOS Devices

An attractive alternative to the CCD readout is coordinative addressing with CMOS switches. In particular, silicon fabrication advances now permit the implementation of CMOS transistor structures that are considerably smaller than the wavelength of visible light and which have enabled the practical integration of multiple transistors within a single pixel. The configuration of CCD devices requires specialized processing, unlike CMOS imagers, which can be built on fabrication lines designed for commercial microprocessors. CMOS have the advantage that existing foundries, intended for application to specific integrated circuits (ASICs), can be readily used by adapting their design rules. Design rules of 7-nm are currently in production, with pre-production runs of 5-nm design rules. As a result of such fine design rules, more functionality has been designed into the unit cells of multiplexers with smaller unit cells, leading to large array sizes. Figure 4.2 shows the timelines for minimum circuit features and the resulting CCD, IR FPA, and CMOS visible imager sizes with respect to the number of imaging pixels. Along the horizontal axis is also a scale depicting the general availability of various MOS and CMOS processes. The ongoing migration to even finer lithography will thus enable the rapid development of CMOS-based imagers having even greater resolution, better image quality, higher levels of integration, and lower overall imaging system cost than CCD-based solutions. At present, CMOS, with minimum features of $\leq 0.1$-$\mu$m, makes monolithic visible CMOS imagers possible, because the denser photolithography allows for low-noise signal extraction and high-performance detection with high optical fill-factor within each pixel. The pixel's architecture is changed to improve resolution by shrinking the pixel size. Figure 4.8 is a roadmap of where CMOS pixel pitch became smaller than CCD due to the described technological developments in 2010 [23]. CMOS imagers are also moving rapidly to large formats and, at present, compete with CCDs for the large-format applications. The silicon wafer production infrastructure, which has put high-performance personal computers into many homes, makes CMOS-based imaging in consumer products, such as video and digital still cameras, widely available.

A typical CMOS multiplexer architecture [Fig. 4.6(b)] consists of fast (column) and slow (row) shift registers at the edges of the active area, and pixels are addressed, one by one, through the selection of a slow register, whereas the fast register scans through a column, and so on. Each image-sensor is connected in parallel to a storage capacitor located in the unit cell. A column of diodes and storage capacitors is selected, one at a time, by a digital horizontal scan register and a row bus is selected by the vertical scan register. Therefore, each pixel can be individually addressed.

CMOS-based imagers use active and passive pixels [13,17] as shown, in simplified form, in Fig. 4.5(b). In comparison with passive pixel sensors (PPSs), active pixel sensors (APSs), apart from read functions, exploit some form of amplification at each pixel. The PPS consists of three transistors: a reset FET, a selective switch, and a source follower (SF) for driving the signal

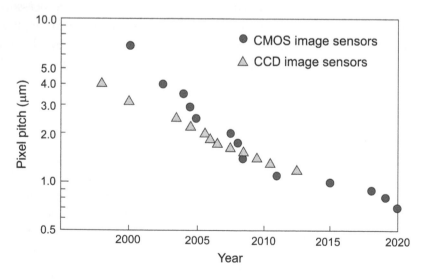

FIGURE 4.8  A roadmap of CMOS pixel pitch development.

onto the column bus. As a result, circuit overhead is low and the optical collection efficiency [fill factor (FF)] is high, even for a monolithic device. Micro-lenses, typically used in CCD and CMOS APS imagers for visible application, concentrate the incoming light into the photosensitive region, where they are accurately deposited over each pixel (Fig. 4.9). In the case of the per-pixel electronics, the area available in the pixel for the detector is reduced, so FF is often limited to 30 to 60 percent. When the FF is low and micro-lenses are not used, the light falling elsewhere is either lost or, in some cases, creates artifacts in the imagery by generating electrical currents in the active circuitry. Unfortunately, micro-lenses are less effective when used in low f/# imaging systems, and may not be appropriate for all applications.

Although a micro-lens improves the pixel sensitivity, there are many design challenges associated with the application of micron-scale lens arrays. One key issue is crosstalk, which degrades spatial resolution, color separation, and overall sensitivity. Reducing crosstalk in small pixels has become one of the most difficult and time-consuming tasks in sensor design. At pixel dimensions below 1 μm, it becomes more difficult in the frontside metallization of a CMOS sensor, and advanced light-guiding structures need to be used to reduce crosstalk.

In the APS, three of the metal-oxide-semiconductor field-effect transistors (MOSFETs) have the same function as in PPS. The fourth transistor works as a transfer gate that moves charge from the photodiode to the floating diffusion. Usually, both pixels operate in rolling shutter mode. The APS is capable of performing correlated double sampling (CDS) to eliminate the reset noise (kTC noise) and the pixel offsets. The PPS can only be used with noncorrelated double sampling, which is sufficient to reduce the pixel-to-pixel offsets, but does not eliminate the temporal noise (temporal noise can be addressed by other methods). Adding additional components (like soft reset or tapered

Micro-lens array

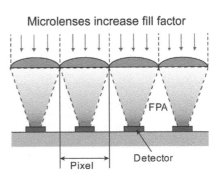

FIGURE 4.9  Micrograph and cross-sectional drawing of micro-lensed FPA.

reset) however, reduces the FF of monolithic imagers [21]. The MOSFETs incorporated in each pixel for readout are optically dead. CMOS sensors also require several metal layers to interconnect MOSFETs. The buses are stacked and interleaved above the pixel, producing an "optical tunnel" through which incoming photons must pass. In addition, most CMOS imagers are frontside illuminated. This limits the visible sensitivity in the red because of a relatively shallow absorption material. In comparison, CCD pixels are constructed so that the entire pixel is sensitive, with a 100% FF.

Figure 4.10 compares the principles of CCD and CMOS sensors. Both detector technologies use a photosensor to generate and separate the

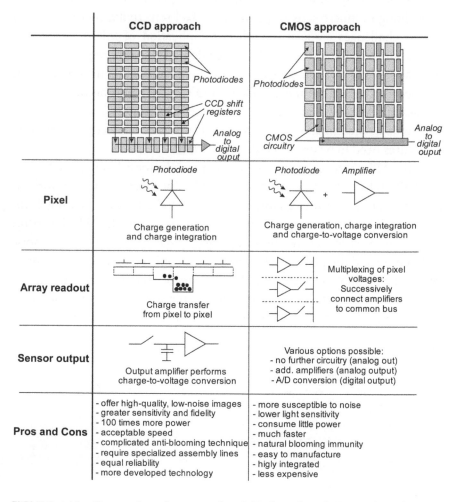

FIGURE 4.10 Comparison between the CCD-based and CMOS-based image sensor approaches.

charges in the pixel. Beyond that, however, the two sensor schemes differ significantly. During CCD readout, the collected charge is shifted from pixel to pixel all the way to the perimeter. Finally, all charges are sequentially pushed to one common location (floating diffusion), and a single amplifier generates the corresponding output voltages. On the other hand, CMOS detectors have an independent amplifier in each pixel (APS). The amplifier converts the integrated charge into a voltage and thus eliminates the need to transfer charge from pixel to pixel. The voltages are multiplexed onto a common bus line, using integrated CMOS switches. Analog and digital sensor outputs are possible by implementing either a video output amplifier or an analog-to-digital (A/D) converter on the chip.

The processing technology for CMOS is typically two to three times less complex than standard CCD technology. In comparison with CCDs, the CMOS multiplexers exhibit important advantages due to high circuit density, fewer drive voltages, fewer clocks, much lower voltages (low power consumption), and packing density compatible with many more special functions, and lower cost for both digital video and still camera applications. The minimum theoretical read noise of a CCD is limited in large imagers by the output amplifier's thermal noise after CDS is applied in off-chip support circuits. The alternative CMOS paradigm offers lower temporal noise because the relevant noise bandwidth is fundamentally several orders of magnitude smaller and better matches the signal bandwidth. Whereas CCD sensitivity is constrained by the limited design space involving the sense node and the output buffer, CMOS sensitivity is limited by only the desired dynamic range and operating voltage. CMOS-based imagers also offer practical advantages with respect to on-chip integration of camera functions, including command and control electronics, digitization, and image processing. CMOS is now suitable also for TDI-type multiplexers because of the availability from foundries of design rules lower than 1.0-µm, more uniform electrical characteristics, and lower noise figures.

## 4.2 Hybrid Arrays

In the case of hybrid technology, we can optimize the detector material and multiplexer independently. Other advantages of hybrid-packaged FPAs are near-100% fill factors and increased signal-processing area on the multiplexer chip. Photodiodes, with their very low power dissipation, inherently high impedance, negligible $1/f$ noise, and easy multiplexing *via* the ROIC, can be assembled in 2D arrays containing a very large number of pixels, limited only by existing technologies. Photodiodes can be reverse-biased for even higher impedance, and can therefore better match electrically with compact low-noise silicon readout preamplifier circuits. The photo-response of photodiodes remains linear for significantly higher photon flux levels than that of photoconductors, primarily because of higher doping levels in the photodiode absorber layer and because the photogenerated carriers are collected rapidly by the junction.

Development of hybrid-packaging technology began in the late 1970's [24] and took the next decade to reach volume production. In the early 1990's, fully 2D imaging arrays provided a means for staring sensor systems to enter the production stage. In the hybrid architecture, indium bump bonding, with readout electronics, provides for multiplexing the signals from thousands or millions of pixels onto a few output lines, greatly simplifying the interface between the vacuum-enclosed cryogenic sensor and the system electronics.

Although focal plane array imagers are very common in our lives, they are quite complex to fabricate. Depending on the array architecture, the process can include over 150 individual fabrication steps. The hybridization process involves flip-chip indium bonding between the "top" surfaces of the ROIC and the detector array. The indium bond must be uniform, between each sensing pixel and its corresponding read-out element, in order to ensure high-quality imaging. After hybridization, a backside-thinning process is usually performed to reduce the amount of substrate absorption. The edges of the gap between the ROIC and the FPA can be sealed with low-viscosity epoxy before the substrate is mechanically thinned down to several microns. Some advanced FPA fabrication processes involve complete removal of the substrate material.

Innovations to and progress in FPA fabrication are dependent on adjustments to the material growth parameters. Usually, in-house growth has provided manufacturers with the ability to maintain the highest-quality material, and to customize the layer structures for multiple applications. For example, since HgCdTe material is critical to many principal product lines, and comparable material is not available externally, most of the global manufacturers continue to supply their own wafers. Figure 4.11 shows process flow for integrated infrared FPA manufacturing. As is shown, boule growth starts with the raw materials, i.e., polycrystalline components. In the case of the HgCdTe FPA process, polycrystalline ultrapure CdTe and ZnTe binary compounds are loaded into a carbon-coated quartz crucible. The crucible is mounted into an evacuated quartz ampoule, which is placed in a cylindrical furnace. Large-crystal CdZnTe boules are produced by mixing and melting the ingredients, followed by recrystallizing, using the vertical gradient freeze method. Their standard diameters reach 125 mm. The boule substrate material is then sawn into slices, diced into squares, and polished to prepare the surface for epitaxial growth. Typical substrate sizes up to 8 cm×8 cm have been produced. The HgCdTe layers are usually grown on top of the substrate by MBE or MOCVD. In the case of MOCVD epitaxial technology, large-size GaAs substrates are also used. The selection of substrate depends on the specific application. The entire growth procedure is automated, with each step being programmed in advance.

After growing the detector epitaxial structures, the wafers are nondestructively evaluated against multiple quality specifications. They are then conveyed to the array processing line, where the sensing elements (pixels) are formed by photolithographic steps, including mesa etching, surface passivation, metal contact deposition, and indium bump formation. After wafer dicing, the FPAs are ready for mating to the ROICs. The ROIC branch of the process is shown in the lower right of Fig. 4.11. For each pixel

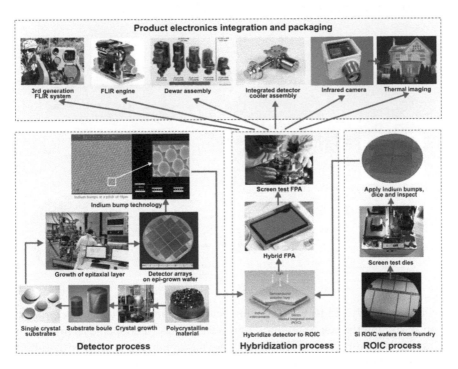

FIGURE 4.11   Process flow for integrated infrared FPA manufacturing.

on the detector array, there is a corresponding unit cell on the ROIC to collect the photocurrent and process the signal. Each design is delivered to a silicon foundry for fabrication. Next, the ROIC wafers are diced and are ready for mating with the FPA. The most advanced flip-chip bonders, utilizing laser alignment and submicron-scale motion control, bring the two chips together (see the center of Fig. 4.11). At present, FPAs with a pixel pitch size below 10-μm are aligned and hybridized with high yield. Each FPA, with attached ROIC, is tested, according to a defined protocol, and is installed in a sensor module. Finally, associated packaging and electronics are designed and assembled to complete the integrated manufacturing process.

Detector FPA has revolutionized many kinds of imaging, from gamma rays to the infrared and even radio waves. More general information about background, history, present stage of technology, and trends can be found e.g. [4,25]. Information about the assemblies and applications can be found at different vendor websites.

Different hybridization approaches are in use today. The most popular is flip-chip interconnect, using bump bond (see Fig. 4.12a,c). In this approach, indium bumps are formed on both the detector array and the

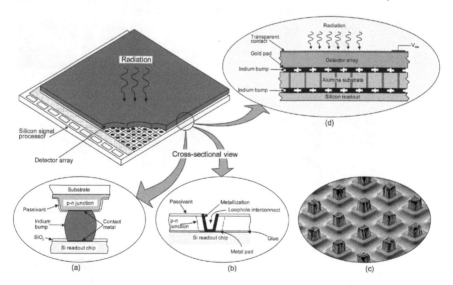

FIGURE 4.12 Hybrid IR FPA interconnect techniques between a detector array and silicon multiplexer: (a) indium bump technique, (b) loophole technique, (c) scanning electron micrograph (SEM) shows mesa photodiode array with indium bumps, and (d) layered-hybrid design suitable for large format far IR and sub-mm arrays.

ROIC chip. The array and the ROIC are aligned and force is applied to cause the indium bumps to cold-weld together. In the other approach, indium bumps are formed only on the ROIC; the detector array is brought into alignment and into proximity with the ROIC, the temperature is raised to cause the indium to melt, and contact is made by reflow.

Large focal plane arrays with fine-pitched bumps may require substantial force to achieve reliable interconnects, which creates the potential for damaging delicate semiconductor detector layers. Also, there is always a risk of misalignment associated with the joining process and "hybrid slip". Some of these issues can be resolved by using a fusion bonding (or direct bond) process, which usually involves a room temperature alignment of wafers *via* van der Waals interactions, followed by annealing, that creates permanent covalent interfacial bonds between upper and lower wafers. The direct bond process requires less than 10 pounds of force (in comparison with > 1000 pounds for the indium-based process) to mate the detector and the ROIC. This eliminates misalignment errors and many potential problems associated with compression bonder (hybrid slip, layer damage, bump separation, etc.). Fusion bonding is used, particularly in production of very large arrays up to 10 k × 12 k format and thin active layers (<10-μm). This approach is beneficial for devices with enhanced blue and UV performance, and modulation transfer function (MTF). The drawback of this method is its extreme sensitivity to surface flatness, roughness, and particles.

Recently, an alternative technology, based upon adhesive bonding between the active detector layer and the support wafer, has been introduced. It provides low temperature bonding and alignment accuracies in the 1-2 μm range. However, the adhesive has to be carefully selected and tested for mechanical rigidity, temperature stability, and outgassing, and inter-via connections must be made after bonding and thinning.

Infrared hybrid FPA detectors and multiplexers are also fabricated using loophole interconnection (Fig. 4.12b) [26,27]. In this case, the detector and the multiplexer chips are glued together to form a single chip before detector fabrication. The photovoltaic detector is formed by ion implantation, and loopholes are drilled by ion-milling and electrical interconnection between each detector and its corresponding input circuit is made through a small hole formed in each detector. The junctions are connected down to the silicon circuit by cutting the fine holes, a few μm in diameter, through the junctions by ion milling, and then the holes are backfilled with metallization. A similar type of hybrid technology called

VIP™ (vertically integrated photodiode) was reported by DRS Infrared Technologies (former Texas Instruments) [28,29].

It is difficult to make small pixel pitches (below 10-μm) using the bump-bonding interconnect technique, especially when high yield and 100% pixel operability are required. A new facility gives a 3D integration process using wafer bonding, where materials such as Si and InP have been monolithically integrated with pixels size down to 6-μm [30,31]. Figure 4.13 compares three methods of vertically interconnected circuit layers: (a) bump bond, (b) insulated through-silicon vias, and (c) Lincoln Laboratory's silicon-on-insulator (SOI)-based via. The Lincoln integration method enables the dense vertical interconnection of multiple circuit layers and is capable of achieving far smaller pixel sizes than is possible with bump bonding.

The detector array can be illuminated from either the frontside (with the photons passing through the transparent silicon multiplexer) or the backside (with photons passing through the transparent detector array substrate). In general, the latter approach is more advantageous, as the multiplexer will typically have areas of metallization and other opaque regions, which can reduce the effective optical area of the structure. The epoxy is flowed into the space between the readout and the detectors to increase the bonding strength. In the case of backside-detector-illumination, transparent substrates are required. When using opaque materials, substrates must be thinned to below 10 μm to obtain sufficient quantum efficiencies and to reduce crosstalk. In some cases, the substrates are completely removed. In the "direct" backside-illuminated configuration, both the detector array and the silicon ROIC chip are bump mounted side-by-side onto a common circuit board. The "indirect" configuration allows the unit cell area in the

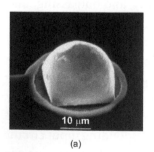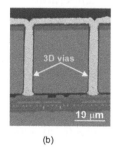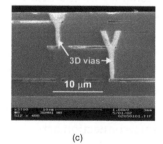

(a)                    (b)                    (c)

FIGURE 4.13 Approaches to 3D integration: (a) bump bond used to flip-chip interconnect two circuit layers, (b) two-layer stack with insulated vias through thinned bulk Si, (c) two-layer stack using Lincoln's SOI-based vias.

silicon ROIC to be larger than the detector area and is usually employed for small scanning FPAs, where stray capacitance is not an issue.

Readout circuit wafers are processed in standard commercial foundries and can be constrained in size by the die-size limits of the photolithography step and repeat printers. Because of field size limitations in those photography systems, CMOS imager chip sizes must currently be limited to standard lithographic field sizes of less than 32×26 mm for submicron lithography. To build larger sensor arrays, a new photolithographic technique called *stitching* can be used to fabricate detector arrays larger than the reticle field of photolithographic steppers. The large array is divided into smaller sub-blocks. Later, the complete sensor chips are stitched together from the building blocks in the reticle. Each block can be photocomposed on the wafer by multiple exposures at appropriate locations. Single blocks of the detector array are exposed at one time, as the optical system allows shuttering, or selectively expose only the desired section of the reticle

It should be noted that stitching creates a seamless detector array, as opposed to an assembly of closely butted sub-arrays [32,33]. The butting technique is commonly used in the fabrication of very large format sensor arrays due to the limited size of substrate wafers.

The traditional hybrid systems are typically limited to 1-megapixel arrays, due to small detector wafer size and low throughput. The size of pixel pitch below 5-μm is limited by the hybridization process, as the solder bumps need sufficient volume for reliable bonding, which, in turn, is limited by the achievable aspect ratio and pixel spacing. With a thin-film active layer integrated monolithically directly on top of the readout circuit, sub-micron pixel sizes can be achieved. For example, using amorphous colloidal quantum dot (CQD) layers as the detector's active region permits fabrication of devices directly onto ROIC substrates, as shown in Fig. 4.14, with no restrictions on pixel or array size and with a day-cycle

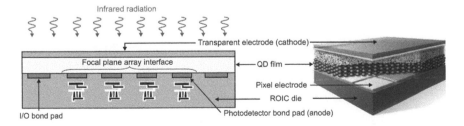

FIGURE 4.14   Infrared monolithic array structure based on colloidal-quantum-dots.

of production. In addition, the monolithic integration of CQD detectors into ROIC does not require any hybridization step. Individual pixels are defined by the area of the metal pads arranged on the top of the ROIC surface [34]. To synthesize colloidal nanocrystals, wet chemistry techniques are used. Reagents are injected into a flask and the desirable shape and size of nanocrystals are obtained by the control of reagent concentrations, ligand selection, and temperature. This so-called top-surface photodetector offers a 100% fill-factor and is compatible with postprocessing at the top of CMOS electronics.

## 4.3 INFRARED FPA PERFORMANCE CONSIDERATIONS

The configuration of the thermal imaging system is shown in Fig. 4.15. On the graph shown in this figure, $A_s$ and $A_d$ are the surfaces of the object and the detector, respectively, $r$ is the distance from the object to the lens (system optics), $A_{ap} = \pi D^2/4$ is the surface, and $D$ the diameter of the lens (aperture, entrance-pupil). The detector is placed in the focal plane of the system in the distance $\approx f$ to the entrance pupil. The optic's system is characterized by the so-called $F$-number, i.e., $f/\# = f/D$.

For arrays, the relevant figure of merit for determining the ultimate performance is not the detectivity, $D^*$, but the noise-equivalent difference temperature (*NEDT*) and the modulation transfer function (*MTF*). These are considered to be the primary performance metrics for thermal imaging systems, namely thermal sensitivity and spatial resolution. Thermal sensitivity is concerned with the minimum temperature difference that can be discerned above the noise level. The *MTF* concerns the spatial resolution and answers the question of how small an object can be imaged by

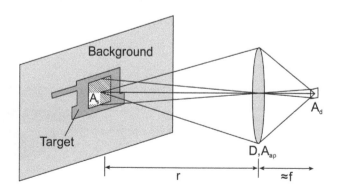

FIGURE 4.15   Thermal imaging system configuration.

the system. The general approach to system performance is given by Lloyd in his fundamental monograph [35].

### 4.3.1 Modulation Transfer Function

Modulation transfer function (*MTF*) is the ability of an imaging system to faithfully image a given object, and quantifies the ability of the system to resolve or transfer spatial frequencies [36]. Consider a bar pattern with a cross-section of each bar being a sine wave. Since the image of a sine wave light distribution is always a sine wave, the image is always a sine wave independent of the other effects in the imaging system, such as aberrations.

Usually, imaging systems have no difficulty in reproducing the bar pattern when the bar pattern is closely spaced. However, an imaging system reaches its limit when the features of the bar pattern get closer and closer together. When the imaging system reaches this limit, the contrast or the modulation (*M*) is defined as:

$$M = \frac{E_{max} - E_{min}}{E_{max} + E_{min}}, \qquad (4.1)$$

where *E* is the irradiance. Once the modulation of an image is measured experimentally, the *MTF* of the imaging system can be calculated for that spatial frequency, using

$$MTF = \frac{M_{image}}{M_{object}}. \qquad (4.2)$$

The system *MTF* is dominated by the optics, detector, and display *MTF*s and can be cascaded by simply multiplying the *MTF* components to obtain the *MTF* of the combination. In spatial frequency terms, the *MTF* of an imaging system at a particular operating wavelength is dominated by limits set by the size of the detector and the aperture of the optics. More details about this issue are given in Section 4.5.

### 4.3.2 Noise Equivalent Difference Temperature

The *NEDT* of a detector represents the temperature change, for incident radiation, that gives an output signal equal to the root-mean-square (rms) noise level. Although normally thought of as a system parameter, the detector *NEDT* and the system *NEDT* are the same, except for system losses. *NEDT* is defined as:

$$NEDT = \frac{V_n \left( \partial T / \partial \Phi \right)}{\left( \partial V_s / \partial \Phi \right)} = V_n \frac{\Delta T}{\Delta V_s}, \tag{4.3}$$

where $V_n$ is the rms noise, $\Phi$ is the spectral photon flux density (photons/cm²/s) incident on a focal plane, and $\Delta V_s$ is the signal measured for the temperature difference, $\Delta T$.

We further follow after Kinch [37] to obtain useful equations for noise equivalent irradiance (NEI) and NEDT, used for estimation of detector performance.

In modern IR focal plane arrays (FPAs), the current generated in a biased photon detector is integrated onto a capacitive node with a carrier well capacity of $N_w$. Values for $N_w$ are typically in the range of $1 \times 10^6$ to $1 \times 10^7$ electrons for a 15-μm pixel design with available node capacities for current CMOS readout IC designs.

For an ideal system, in the absence of excess noise, the detection limit of the node is achieved when a minimum detectable signal flux, $\Delta \Phi$ creates a signal equal to the shot noise on the node:

$$\Delta \Phi \eta A_d \tau_{int} = \sqrt{N_w} = \sqrt{\frac{(J_d + J_\Phi) A_d \tau_{int}}{q}}, \tag{4.4}$$

where $\eta$ is the detector collection efficiency, $A_d$ is the detector area, $\tau_{int}$ is the integration time, $N_w$ is the well capacity of the readout, $J_d$ is the detector dark current, and $J_\phi = \eta \Phi_B A$ is the background flux current.

Another critical parameter is connected with NEDT, the so-called noise equivalent flux (NE$\Delta\Phi$). This parameter is defined for spectral regions in which the thermal background flux does not dominate. By equating the minimum detectable signal to the integrated current noise, we have:

$$\eta \Phi_s A_d \tau_{int} = \sqrt{\frac{(J_d + J_\Phi) A_d \tau_{int}}{q}}, \tag{4.5}$$

giving

$$NE\Delta\Phi = \frac{1}{\eta} \sqrt{\frac{J_d + J_\Phi}{q A_d \tau_{int}}}. \tag{4.6}$$

This can be converted to a noise equivalent irradiance (NEI), which is defined as the minimum observable flux power incident on the system aperture, by renormalizing the incident flux density on the detector to the system aperture area, $A_{opt}$. The $NEI$ is given by:

$$NEI = NE\Delta\Phi \frac{A_d h\nu}{A_{apt}},$$ (4.7)

where the monochromatic radiation of energy, $h\nu$ is assumed.

$NEI$ [photons/(cm²/s)] noise equivalent irradiance is the signal flux level at which the signal produces the same output as the noise present in the detector. This unit is useful because it directly gives the photon flux above which the detector will be photon noise-limited.

For high-background-flux conditions, the signal flux can be defined as $\Delta\Phi = \Delta T(d\Phi_B/dT)$. Thus, for shot noise, substituting in Eq. (4.4), we have:

$$\eta\Delta T \frac{d\Phi_B}{dT} = \sqrt{\frac{(J_d + J_\Phi) A_d \tau_{int}}{q}}.$$ (4.8)

Finally, after some re-arrangement:

$$NEDT = \frac{1 + (J_d/J_\Phi)}{\sqrt{N_w C}},$$ (4.9)

where $C = (d\Phi_B/dT)/\Phi_B$ is the scene contrast through the optics. The scene contrast in the MWIR band at 300 K is 3.5–4% compared to 1.7% for the LWIR band. In deriving the last equation, it was assumed that the optics transmission is unity, and that the cold shield of the detector is not contributing flux. This is reasonable at low detector temperatures but not at higher operating temperatures. At higher temperatures, the scene contrast is defined in terms of the signal flux coming through the optics, whereas the flux current is defined by the total flux through the optics and the flux from the cold shield.

According to Eq. (4.9), if the value of $J_d/J_\Phi$ ratio increases and/or the value of $\eta$ decreases, more prolonged integration time and faster speed of the optics are required. Thus, less-efficient detectors can be utilized in faster optics and slower frame rate systems. On the other hand, the efficient pixel detectors are characterized by low dark current density and high quantum efficiency and can be used in thermal imagers with slow optics and faster frame rates.

The $NEDT$ also characterizes the thermal sensitivity of an infrared system, i.e., the amount of temperature difference required to produce a unity signal-to-noise ratio. A smaller *NEDT* indicates a better thermal sensitivity. In spite of its widespread use in the infrared literature, it is applied to different systems, under different conditions, and with different meanings [38]. From considerations carried out in Ref. [35], it can be shown that:

$$NEDT = \frac{4(f/\#)^2 \Delta f^{1/2}}{A_d^{1/2}} \left[ \int_{\lambda_a}^{\lambda_b} \frac{\partial M(\lambda)}{\partial T} D^*(\lambda) d\lambda \right]^{-1}, \qquad (4.10)$$

where the spectral range (atmospheric windows) is limited by wavelengths $\lambda_a$ and $\lambda_b$. In the above consideration, both atmosphere and optics transmissions have been assumed to equal 1.

To achieve the greatest sensitivity (i.e., the lowest *NEDT*), the spectral integral in Eq. (4.10) should be maximized. This can be obtained when the peak of the spectral responsivity and the peak of the radiant exitance contrast, $\partial M/\partial T$, coincide. However, the thermal imaging system may not satisfy these conditions because of other constraints, such as atmospheric/obscurant transmittance effects or available detector characteristics. Dependence on the square root of bandwidth is intuitive, since the root-mean-square (rms) noise is proportional to $(\Delta f)^{1/2}$. In addition, better *NEDT* results are obtained from lower *f/#*. A lower *f/#* number results in more flux being captured by the detector, resulting in increased signal-to-noise (SNR) for a given level.

The dependence of *NEDT* on the detector area is critical. The inverse-square-root dependence of *NEDT* on the detector area results from an effect of two terms: increasing rms noise as the square root of the detector area, and proportional increase of the signal voltage to the area of detector. The net result is that $NEDT \propto 1/(A_d)^{1/2}$. Whereas the thermal sensitivity of an imager is better for larger detectors, the spatial resolution is poorer for larger detectors (pixels). Hence, a reasonable compromise between the requirement of high thermal and spatial resolution is necessary. Another parameter, the minimum resolvable difference temperature (*MRDT*), considers both thermal sensitivity and spatial resolution, more appropriate for design.

As Eq. (4.10) shows, improvement of thermal resolution, without detrimental effects on spatial resolution, may be achieved by:

- A decrease of detector area combined with a corresponding decrease of the optics *f/#*,

- Improved detector performance, and

- An increase in the number of detectors.

An increase in aperture size is undesirable because it increases the size, mass, and price of an IR system. It is more appropriate to use a detector with greater detectivity. This can be achieved by better coupling of the detector with the incident radiation. Another possibility is the use of a multi-elemental sensor, which reduces each element bandwidth proportionally to the number of elements for the same frame rate and other parameters.

The above considerations are valid, assuming that the temporal noise of the detector is the main source of noise. However, this assertion is not true to staring arrays, where the nonuniformity of the detector's response is a significant source of noise. This nonuniformity appears as a fixed pattern noise (spatial noise). It is defined in various ways in the literature; however, the most common definition is that it is the dark signal nonuniformity arising from an electronic source (i.e. other than thermal generation of the dark current), e.g., clock breakthrough or from offset variations in row, column, or pixel amplifiers/switches. So, estimation of IR sensor performance must include a treatment of spatial noise that occurs when FPA nonuniformities cannot be compensated for directly.

Mooney *et al.* [39] have given a comprehensive discussion of the origin of spatial noise. The total noise of a staring array is the composite of the temporal noise and the spatial noise. The spatial noise is the residual nonuniformity $u$ after application of nonuniformity compensation, multiplied by the signal electrons $N$. Photon noise, equal to $N^{1/2}$, is the dominant temporal noise for the high IR background signals for which spatial noise is significant. Then, the total *NEDT* is

$$NEDT_{total} = \frac{\left(N + u^2 N^2\right)^{1/2}}{\partial N / \partial T} = \frac{\left(1/N + u^2\right)^{1/2}}{\left(1/N\right)\left(\partial N / \partial T\right)}, \qquad (4.11)$$

where $\partial N / \partial T$ is the signal change for a 1-K source temperature change. The denominator, $\left(\partial N / \partial T\right)/N$, is the fractional signal change for a 1-K source temperature change. This is the relative scene contrast.

The dependence of the total *NEDT* on detectivity for different residual nonuniformity is plotted in Fig. 4.16 for a 300 K scene temperature and a set of parameters shown as insets in the figure. When the detectivity is

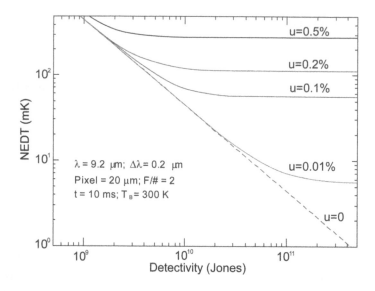

FIGURE 4.16   *NEDT* as a function of detectivity. The effects of nonuniformity are included for $u = 0.01\%$, 0.1%, 0.2%, or 0.5%. Note that, for $D^* > 10^{10}$ cmHz$^{1/2}$/W, detectivity is not the relevant figure of merit.

approaching a value above $10^{10}$ cmHz$^{1/2}$/W, the FPA performance is uniformity limited, prior to correction, and is thus essentially independent of detectivity. An improvement in nonuniformity from 0.1% to 0.01% after correction could lower the *NEDT* from 63 to 6.3 mK.

The nonuniformity value is usually calculated using the standard deviation divided by the mean (the coefficient of variation), counting the number of operable pixels in an array. For a system operating in the LWIR band, the scene contrast is about 2%/K for a change in scene temperature. Thus, to obtain a pixel-to-pixel variation in apparent temperature to less than, e.g., 20 mK, the nonuniformity in response must be less than 0.04%. This is almost impossible to achieve in the uncorrected response of the FPA, so a two-point correction is typically used.

## 4.4  PRESENT STATUS OF INFRARED FPAs

Infrared system performance is highly scenario dependent and requires the designer to account for numerous different factors when specifying detector performance. It means that a good solution for one application may not be as suitable for a different application. In general, detector material is primarily selected on the basis of wavelength of interest, performance criterion, and operating temperature (Fig. 4.17). Although

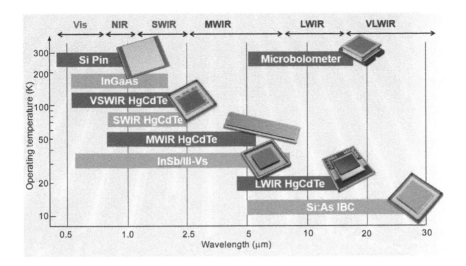

FIGURE 4.17 Detector materials which have the largest interest for infrared detector technology (adapted after Ref. [40]).

efforts have been made to develop monolithic structures using a variety of infrared detector materials (including narrow-bandgap semiconductors) over the past 40 years, only a few have matured to the level of practical use. These included Si, PtSi, and, more recently, microbolometers, PbS, and PbSe. Other infrared material systems (InGaAs, HgCdTe, InSb/III-Vs type-II superlattice, GaAs/AlGaAs QWIP, Si:As BIB) are used in hybrid configurations.

A suitable detector material for near-IR (1.0–1.7-μm) spectral range is Si and InGaAs lattice matched to the InP. Various HgCdTe alloys, usually in photovoltaic configuration, cover from 0.7-μm to over 20-μm. Similar spectral responsivity have InAs/GaSb strained-layer superlattices. Impurity-doped (Sb, As, and Ga) silicon blocked-impurity-band (BIB) detectors operating at very low temperature have a spectral response cutoff in the range of 16- to 30-μm. Ge:Ga photoconductors are the best low-background photon detectors for the wavelength range from 40- to 120-μm, operating at about 2 K.

Tables 4.1 and 4.2 contain a description of representative microbolometer and photon IR FPAs that are commercially available as standard products and/or catalog items from the major manufacturers.

Thermal imaging systems are used first to detect an object and then to identify it. In military circles, "identification" (I) is used along with "detection" (D), and "recognition" (R) as part of the DRI criteria established by

TABLE 4.1 Representative Commercial Uncooled Infrared Bolometer Array

| Company | Bolometer Type | Array Format | Pixel Pitch (μm) | Detector NEDT (mK) (f/1, 20–60 Hz) | Time Constant (ms)/Frame Rate (Hz) |
|---|---|---|---|---|---|
| **L-3** (USA) www.l3t.com | $VO_x$ bolometer | 320×240 | 37.5 | 50 | |
| | a-Si bolometer | 160×120 – 640×480 | 30 | 50 | |
| | a-Si/a-SiGe bolometer | 320×240 – 1024×768 | 17 | 30–50 | |
| **BAE** (USA) www.fairchildimaging.com | $VO_x$ bolometer | 640×480 | 12 | <50 | <15 ms |
| | $VO_x$ bolometer | 1920×1200 | 12 | <50 | |
| **DRS** (USA) www.leonardodrs.com | $VO_x$ bolometer | 320×240 | 25 | <40 | ≤18 ms/60 Hz |
| | $VO_x$ bolometer | 320×240 | 17 | <50 | 60 Hz |
| | $VO_x$ bolometer | 640×480 | 17 | <40 | ≤14 ms/30 Hz |
| | $VO_x$ bolometer | 1024×768 | 17 | <40 | ≤14 ms/30 Hz |
| | $VO_x$ bolometer | 640×512 | 10 | <50 | 60 Hz |
| **Raytheon** (USA) http://www.raytheon.com/ | $VO_x$ bolometer | 320×240, 640×480 | 25 | 30–40 | |
| | $VO_x$ bolometer | 320×240, 640×480 | 17 | 50 | |
| | $VO_x$ bolometer | 1024×480, 2048×1536 | 17 | | |
| **Lynred** (France) www.lynred.com/products | a-Si bolometer | 80×80 | 17 | <100 | 100 Hz |
| | a-Si bolometer | 160×120 | 12 | <60 | <10 ms/60 Hz |
| | a-Si bolometer | 320×240 | 17 | <60 | <10 ms/60 Hz |
| | a-Si bolometer | 384×288 | 17 | <60 | <10 ms/60 Hz |
| | a-Si bolometer | 640×480 | 17 | <50 | <12 ms/120 Hz |
| | a-Si bolometer | 1024×768 | 17 | <50 | <12 ms/120 Hz |
| **SCD** (Israel) www.scd.co.il | $VO_x$ bolometer | 640×480 | 17 | <35 | <18 |
| | $VO_x$ bolometer | 1024×768 | 17 | <35 | <14 |

*(Continued)*

TABLE 4.1 (CONTINUED)  Representative Commercial Uncooled Infrared Bolometer Array

| Company | Bolometer Type | Array Format | Pixel Pitch (μm) | Detector NEDT (mK) (f/1, 20–60 Hz) | Time Constant (ms)/Frame Rate (Hz) |
|---|---|---|---|---|---|
| **FLIR Systems** (USA) | VO$_x$ bolometer | 640×512 | 12 | <60 | 8 ms/60 Hz |
| http://www.flir.com | VO$_x$ bolometer | 320×256 | 12 | <50 | 8 ms/60 Hz |
| **NEC** (Japan) | VO$_x$ bolometer | 480×360 | 23.5 | 25 | |
| http://www.nec.com | VO$_x$ bolometer | 640×480 | 23.5 | <75 | |
| | VO$_x$ bolometer | 640×480 | 12 | 60 | |
| | VO$_x$ bolometer | 320×240 | 23.5 | NEP <100 pW[*] | |
| **DALI** (China) | a-Si bolometer | 640×480 | 17 | <60 | 5 ms |
| www.dalithermal.com | a-Si bolometer | 384×288 | 17 | <60 | 5 ms |
| | a-Si bolometer | 160×120 | 25 | <60 | 5 ms |
| | a-Si bolometer | 640×480 | 20 | 50 | 15 ms |

[*]at 4 THz

TABLE 4.2  Representative IR Hybrid FPAs Offered by Some Major Manufacturers

| Manufacturer/Web Site | Size/Architecture | Pixel Size (μm) | Detector Material | Spectral Range (μm) | Operating Temperature (K) | $D^*(\lambda_p)$ (cmHz$^{1/2}$/W)/ NETD (mK) |
|---|---|---|---|---|---|---|
| **Sensors Unlimited** (USA) | 640×512 | 12.5×12.5 | InGaAs | 0.7–1.7 | 300 | $2.9 \times 10^{13}$ |
| www.sensorsinc.com | 1280×1024 | 12.5×12.5 | InGaAs | 0.4–1.7 | 300 | $2.9 \times 10^{13}$ |
| **Raytheon Vision Systems** | 1024×1024 | 30×30 | InSb | 0.6–5.0 | 50 | |
| (USA) | 2048×2048 (Orion II) | 25×25 | HgCdTe | 0.6–5.0 | 32 | |
| www.raytheon.com/ | 2048×2048 (Virgo-2k) | 20×20 | HgCdTe | 0.8–2.5 | 4–10 | 23 |
| | 2048×2048 | 15×15 | HgCdTe/Si | 3.0–5.0 | 78 | |
| | 1024×1024 | 25×25 | Si:As | 5–28 | 6.7 | |
| | 2048×1024 | 25×25 | Si:As | 5–28 | | |
| **Teledyne Imaging Sensors** | 4096×4096 (H4RG) | 10×10 or 15×15 | HgCdTe | 1.0–1.7 | 120 | |
| (USA) | 4096×4096 (H4RG) | 10×10 or 15×15 | HgCdTe | 1.0–2.5 | 77 | |
| www.teledyne-si.com/ | 4096×4096 (H4RG) | 10×10 or 15×15 | HgCdTe | 1.0–5.4 | 37 | |
| | 2048×2048 (H2RG) | 18×18 | HgCdTe | 1.0–1.7 | 120 | |
| | 2048×2048 (H2RG) | 18×18 | HgCdTe | 1.0–2.5 | 77 | |
| | 2048×2048 (H2RG) | 18×18 | HgCdTe | 1.0–5.4 | 37 | |
| **Lynred** (France) | 640×512 | 15×15 | InGaAs | 0.9–1.7 | 300 | |
| www.lynred.com/products | 320×256 (Mars) | 30×30 | HgCdTe | 7.7–11.7 | 50–75 | <25 |
| | 640×512 (Scorpio) | 15×15 | HgCdTe | 3.7–4.8 | 90 | 18 |
| | 640×512 (Leo) | 15×15 | HgCdTe | 3.7–4.8 | 80 | 20 |
| | 1280×720 (Daphnis) | 10×10 | HgCdTe | 3.7–4.8 | 120 | 20 |
| | 1280×1024 (Jupiter) | 15×15 | HgCdTe | 3.7–4.8 | 110 | 20 |
| | 640×512 (Scorpio) | 15×15 | HgCdTe | 7.7–9.3 | 90 | 22 |

(Continued)

TABLE 4.2 (CONTINUED)   Representative IR Hybrid FPAs Offered by Some Major Manufacturers

| Manufacturer/Web Site | Size/Architecture | Pixel Size (μm) | Detector Material | Spectral Range (μm) | Operating Temperature (K) | $D^\star(\lambda_p)$ (cmHz$^{1/2}$/W)/ NETD (mK) |
|---|---|---|---|---|---|---|
| **Selex** (United Kingdom) www.leonardocompany.com | 320×256 (Saphira) | 24×24 | HgCdTe APD | 0.8–2.5 | up to 110 | 17 |
| | 640×512 (Hawk MW) | 16×16 | HgCdTe | 3.7–4.95 | | |
| | 640×512 (Hawk MW) | 12×12 | HgCdTe | 3.7–4.95 | | 20 |
| | 1280×720 (SuperHawk) | 8×8 | HgCdTe | 3.7–4.95 | 110 | 32 |
| | 640×512 (Hawk LW) | 16×16 | HgCdTe | 8–10 | up to 90 | |
| | 640×512 (CondorII) | 24×24 | HgCdTe | MW/LW(dual) | 80 | 28/28 |
| **IAM** www.aim-ir.com | 640×512 | 15×15 | HgCdTe | 3.4–5.0 | 95–120 | 18 |
| | 1280×1024 | 15×15 | HgCdTe | 3.4–5.0 | 95–120 | 25 |
| | 640×512 | 15×15 | HgCdTe | 7.6–9.0 | 80 | 23 |
| **SCD** www.scd.co.il | 640×512 | 10×10 | InSb | 3.6–4.9 | 77 | <25 |
| | 1280×1024 | 15×15 | InSb | 3–5 | 77 | 20 |
| | 1920×1536 | 10×10 | InSb | 1–5.4 | | <25 |
| | 1280×1024 | 15×15 | InAsSb nBn | 3.6–4.2 | 150 | <28 |
| | 640×512 | 15×15 | InAs/GaSb T2SL | 7.0–9.3 | 77 | 15 |
| **FLIR Systems** (USA) www.flir.com | 640×512 | 15×15 | InGaAs | 0.9–1.7 | 300 | $10^{10}$ ph/cm$^2$s(NEI) |
| | 1024×1024 | 18×18 | InSb | 3–5 | 80 | <25 |
| | 640×512 | 15×15 | InAs/GaSb T2SL | 7.5–12 | 80 | <40 |
| **DRS Technologies** (USA) www.leonardodrs.com | 1280×760 | 6×6 | HgCdTe | 3.4–4.8 | | <30 |
| | 640×480 | 12×12 | HgCdTe | 3.4–4.8 | | 25 |
| | 2048×2048 | 18×18 | Si:As | 5–28 | 7.8 | |
| | 1024×1024 | 25×25 | Si:As | 5–28 | 7.8 | |
| | 2048×2048 | 18×18 | Si:Sb | 5–40 | 7.8 | |

John Johnson in 1958 [41]. This standard was established to define the performance of thermal imaging cameras. According to Johnson's criteria:

- Detection is defined as the ability to distinguish an object from the background,

- Recognition is defined as the ability to classify the object class (animal, human, vehicle, boat …),

- Identification is defined as the ability to describe the object in detail (a man with a hat, a deer, a Jeep …).

The nominal range performance of an infrared camera is calculated for a defined task, a standardized target, and specific environmental conditions. The only standardization available to date is STANAG 4347 [42].

Figure 4.18 is a comparison of various FPAs in detecting, recognizing, and identifying a man-sized target for a canonical tactical sensor ($f$/3, 454 mm focal length, 152 mm aperture, operating at 60 Hz). These ranges were calculated, using the NVTherm model. It should be mentioned that estimations for type-II superlattice FPAs indicate the same range performance as the HgCdTe ternary alloy.

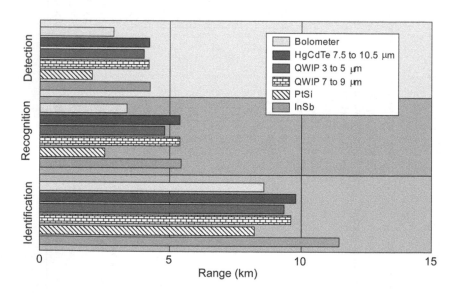

FIGURE 4.18 Comparison DRI ranges for a man-sized target, assuming atmospheric parameters of a mid-latitude (MidLat) summer and rural 23-km visibility (adapted after Ref. [43]).

Typically, identification ranges are between two and three times shorter than detection ranges. To increase ranges, better resolution and greater sensitivity of the infrared systems (and hence the detectors) are required. Further increase in identification ranges can be achieved by using multi-spectral detection to correlate the images at different wavelengths. For that reason, third-generation imagers (Fig. 1.1) are being developed to extend the range of target detection and identification, and to ensure that defense forces maintain a technological advantage in night operations over any opposing force [44].

The unit cell of integrated multicolor FPAs consists of several collocated detectors, each sensitive to a different spectral band (Fig. 4.19). Radiation is incident on the shorter-band detector, with the longer wave radiation passing through to the next detector. Each layer absorbs radiation up to its cutoff, and hence is transparent to the longer wavelengths, which are then collected in subsequent layers. In the case of HgCdTe, this device architecture is realized by placing a longer-wavelength HgCdTe photodiode optically behind a shorter-wavelength photodiode.

Figure 4.20 compares the relative detection and identification ranges modeled for third-generation imagers, using NVESD's (Fort Belvoir, Virginia) NVTherm program. As a range criterion, the standard 70% probability of detection or identification is assumed. Note that the identification range in the MWIR range is almost 70% of the LWIR detection range. For detection, LWIR provides a superior range. In the detection mode, the optical system provides a wide field-of-view (WFOV–*f*/2.5), as the third-generation systems will operate as an on-the-move wide-area step-scanner with automated target recognition (second-generation

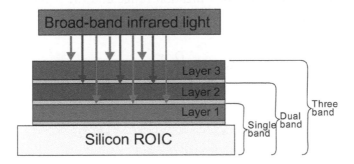

FIGURE 4.19 Structure of a three-color detector pixel. Infrared flux from the first band is absorbed in Layer 3, while longer-wavelength flux is transmitted through the next layers. The thin barriers separate the absorbing bands.

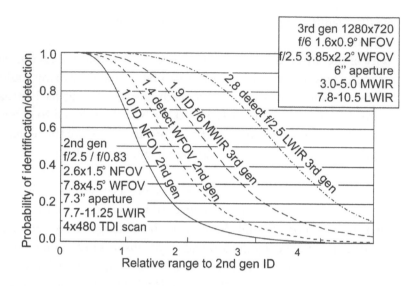

FIGURE 4.20 Comparison of the detection and identification range between current second-generation TDI-scanned LWIR imagers and the LWIR and MWIR bands of third-generation imagers in a 1280×720 format with 20-μm pixels (after Ref. [45]).

systems relay on manual target searching) [45]. MWIR offers greater spatial resolution sensing and has the advantage for long-range identification when used with telephoto optics (NFOV–*f*/6).

## 4.5 TRENDS IN DEVELOPMENT OF INFRARED FPAs

It is well known that detector size, $d$, and F-number ($f/\#$) are primary parameters of infrared systems [46]. These two parameters have a major impact on both detection/identification range and the *NEDT*, since they depend on $F\lambda/d$ [37]:

$$Range = \frac{D\Delta x}{M\lambda}\left(\frac{F\lambda}{d}\right), \qquad (4.12)$$

$$NEDT \approx \frac{2}{C\lambda\left(\eta\Phi_B^{2\pi}\tau_{int}\right)^{1/2}}\left(\frac{F\lambda}{d}\right), \qquad (4.13)$$

where $\lambda$ is the wavelength, $D$ is the aperture, $M$ is the number of pixels required to identify a target, $\Delta x$, $C$ is the scene contrast, $\eta$ is the detector collection efficiency, $\Phi_B^{2\pi}$ is the background flux into a $2\pi$ FOV, and $\tau_{int}$ is the integration time. The above two equations indicate that the parameter space defined by $F\lambda$ and $d$ can be utilized in the optimum design of any IR system.

Most military systems today have the classical view presented in Fig. 4.21 [47], where detector size ranged from 10- to 50-μm. For long-range identification systems, high $f/\#$ optics are used (for a given aperture) to reduce the detector angular subtense. On the other hand, wide field-of-view (WFOV) systems are typically low $f/\#$ systems with short focal lengths as the focal plane had to be spread over wide angles. In recently published papers, it has been shown that long-range identification does not need to be limited to high $f/\#$ systems and that very small detectors can achieve high performances with a smaller package [48–50].

The detection range of many uncooled IR imaging systems is limited by pixel resolution rather than sensitivity. Figure 4.22 presents a trade-off analysis of the detection range and sensor optics for a thermal weapon sight, using the NVESD NVTherm IP model, assuming a detector sensitivity of 35 mK $NETD$ ($f/1$, 30 Hz) for the 25-, 17-, and 12-μm pitch pixel

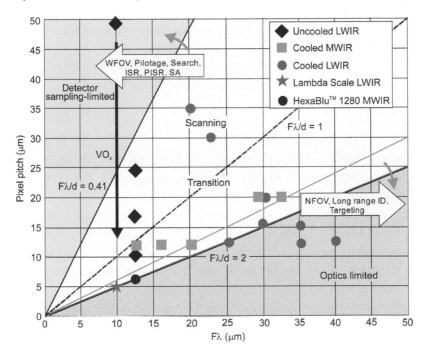

FIGURE 4.21 $F\lambda/d$ space for infrared system design. Straight lines represent constant $NEDT$. There are an infinite number of combinations that provide the same range (adapted after Ref. [47]).

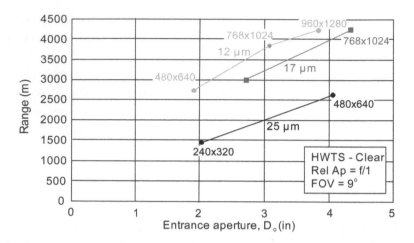

FIGURE 4.22  Calculated detection range as a function of sensor optics and detector pixel size and format, using NVESD NVTherm IP modeling, assuming a 35 mK *NETD* (*f*/1, 30 Hz) for all detectors (adapted after Ref. [51]).

of uncooled FPAs. The advantages of a small pixel pitch and large-format FPAs are obvious. By switching to a smaller pitch and larger format detectors, the detection range of a weapon sight increases significantly with a fixed optic.

The fundamental limit of pixel size is determined by diffraction. The size of diffraction-limited optical spot or Airy disk is given by:

$$d = 2.44\lambda F \qquad (4.14)$$

where $d$ is the diameter of the spot, and $\lambda$ is the wavelength. The spot size for *f*/# ranging from *f*/1 to *f*/10 are shown in Fig. 4.23. For typical *f*/2.0 optics at 4-μm wavelength, the spot size is 20 μm.

It is generally interesting to investigate pixel scaling beyond the diffraction limit, using wavelength- and even subwavelength-scale optics, that are enabled by modern nanofabrication (diffraction-limited pixel size is still relatively large compared with the feature size that can be achieved with state-of-the-art nanofabrication approaches).

FPAs of 1 cm² still dominate the IR market, while pixel pitch has decreased to below 15-μm over the past few years, now reaching 12-μm [52], 10-μm [53,54], 8-μm [55], and even 5-μm in test devices [56,57]. Leonardo DRS offers HexaBlu™ camera modules with a 6-μm pixel pitch, 1280×960 HgCdTe MWIR FPA technology that leverages DRS's proprietary HDVIP® [High Density Vertically Integrated Photodiode, similar

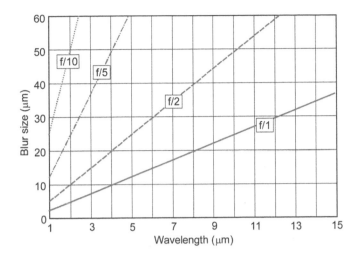

FIGURE 4.23  Optics diffraction limit. The spot size of a diffraction-limited optical system is the Airy disk diameter.

FIGURE 4.24  HexaBlu™ cryo-cooled MW thermal camera module (after Ref. [58]).

to that shown in Fig. 4.12(b)] and detection in a small Integrated Dewar Cooler Assembly (IDCA) (Fig. 4.24). This new pixel design is enabled by HexaBlu's miniature form factor, weighing in at under 295 g and displacing just 80 cm³. This trend toward small IDCA is expected to continue. Systems operating at shorter wavelengths are more likely to benefit from small pixel sizes because of the smaller diffraction-limited spot size. Diffraction-limited optics with low *F*-numbers (e.g., *f*/1) could benefit from pixels of the order of one wavelength across. Over-sampling the diffractive spot may provide some additional resolution for smaller pixels, but this saturates quickly as the pixel size is decreased. Pixel reduction is mandatory to also achieve cost reduction of a system (reduction of the optics diameter, dewar size, and weight, together with the power, and increased reliability). In addition, smaller detectors provide better resolution [50]. Reduction of the focal plane proportionately to the detector size has not changed the detector field-of-view, so that, in the optics-limited region, smaller detectors have no effect on the system's spatial resolution.

Figure 4.25 shows the influence of pixel shrinkage on format enlargement of Lynred's IR arrays [59]. A catalogue of detectors with pixel pitch of 15-µm [Epsilon (384×288), Scorpio (640×512), and Jupiter (1280×1014)] is compared with the Daphnis 10-µm product family.

Shrinking of pixels decreases system size, weight, and power consumption (SWaP), in consequence reducing the system's cost and increasing the operating temperature in HOT detectors (see Section 3.3). Smaller pixels enhance the value of proposition of the imaging systems and their functionality. An example of their capability is illustrated in Fig. 4.26 [47], which shows DRS's production 640×480 FPAs with pixel size from 20-µm to 12-µm. Additional improvement in SWaP is achieved under HOT

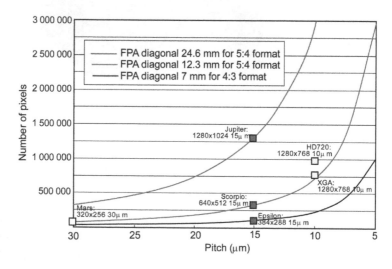

FIGURE 4.25   Number of pixels versus pitch size for Lynred IR FPAs (adapted after Ref. [53]).

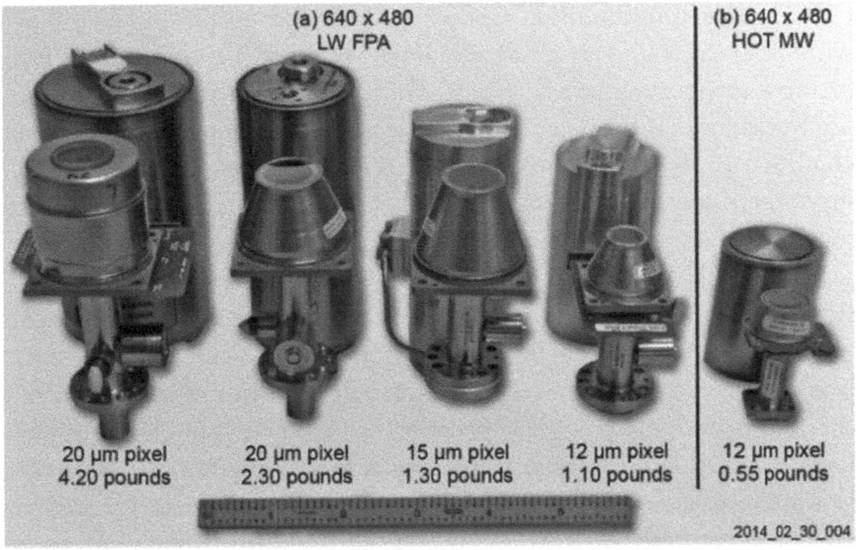

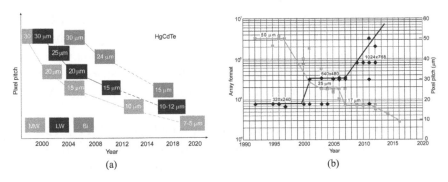

**FIGURE 4.26** Progressive reduction of DRS's 640×480 LWIR package as (a) a function of pixel pitch and (b) increasing of operating temperature (after Ref. [47]).

**FIGURE 4.27** Pixel pitch for (a) HgCdTe photodiodes, and for (b) amorphous silicon microbolometers have continued to decrease due to technological advancements (adapted after Refs. [60, 61].

conditions. Smaller pitches are scheduled in the short term (see Fig. 4.27a) [60]. A similar tendency has been observed in the case of microbolometers (see Fig. 4.27b) [61].

The detector-limited region occurs when $F\lambda/d \leq 0.41$ and the optics-limited region occurs when $F\lambda/d \geq 2$ (Fig. 4.21). When $F\lambda/d = 0.41$, the Airy disk is equal to the detector size. A transition in the region $0.41 \leq F\lambda/d \leq 2.0$ is large and represents a change from detector-limited to optics-limited performance. The condition $F\lambda/d = 2$ is equivalent to placing 4.88 pixels within the Rayleigh blur circle. The lines presented a constant $F\lambda/d$, indicating a constant range and *NEDT*. For a given aperture, $D$, and operating wavelength, $\lambda$, the detection range is given by the optimum resolution condition $F\lambda/d = 2$ and a minimum *NEDT* for a given $\tau_{int}$ [see Eq. (4.13)]. From these considerations, the system $f/\#$ should be fixed to the pixel size to predict the potential limiting performance of IR systems.

Figure 4.21 also includes experimental data points for various classes of thermal imaging systems that have been produced at DRS Technologies, including both uncooled thermal imagers and cooled photon imagers. The earliest uncooled imagers, fabricated at the beginning of the 1990s [barium strontium titanate (BST) dielectric bolometers and $VO_x$ microbolometers], had large pixels of approximately 50-μm pitch and fast optics to achieve useful system sensitivities. With decreasing detector size, the relative apertures remained around $f/1$. As is shown in Fig. 4.21, as the pixel dimensions decreased over time, the uncooled systems steadily progressed from the "detector-limited" regime to the "optics-limited" ones. However, they are still far from the ultimate range capacity for $f/1$ optics.

The cooled thermal imagers include early LWIR scanning systems and modern staring systems, operating under both MWIR and LWIR bands. It has been shown that LWIR imaging systems typically approach $F\lambda/d = 2$ conditions, whereas, for MWIR systems, values of $F\lambda/d$ less than 2 are typically employed, with lower available photon flux making it difficult to maintain system sensitivity. Only the DRS camera (HexaBlu™ MWIR) approaches this limit.

The system *MTF* can be cascaded by simply multiplying $MTF_{opt} \times MTF_{det}$. Figure 4.28 summarizes different behavior of the system *MTF*. The transition region can be further split by setting the optics cutoff frequency to equal the detector cutoff frequency, resulting in an $F\lambda/d = 1.0$ [62]. When $F\lambda/d = 1.0$, the spot size equals 2.44 times the size of the pixel. The "optics-dominated" region lies between the diffraction-limited region and this curve, whereas the "detector-dominated" region is located between this curve and the detector-limited curve. In the optics-dominated region, changes to the optics have a greater impact on the system *MTF* than the detector, and similarly for the detector-dominated region. Historically, most systems have been designed to have a resulting optics blur (to include aberrations) of less than 2.5 pixels ($\sim F\lambda/d < 1.0$). This is, of course, very dependent on the application and range requirements.

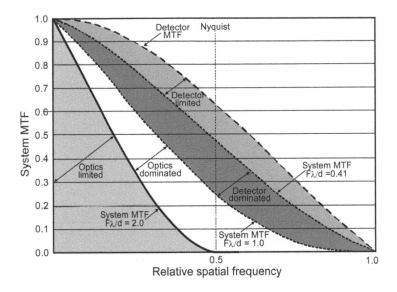

FIGURE 4.28 System *MTF* curves illustrating the different regions with the design space for various *Fλ/d* conditions. Spatial frequencies are normalized to the detector cutoff (adapted after Ref. [62]).

Table 4.3 provides the required *f/#* for $F\lambda/d = 2$ for various detector sizes. As is shown, with *f/*1 optics, the smallest useful detector size is 2-µm in the MWIR and 5-µm in the LWIR. With more realistic *f/*1.2 optics, the smallest useful detector size is 3-µm in the MWIR and 6-µm in the LWIR.

Host and Driggers [49] have considered the influence of optics infrared design on range approximation, which is illustrated in Fig. 4.29. When the IR system is detector-limited, decreasing the detector size has a dramatic effect on the range. On the other size, in the optics-limited region, decreasing the detector size has minimal effect on range performance. The acquisition range is reduced when atmospheric transmission and the *NEDT* are included.

As has been indicated by Kinch [37,63], challenges that must be addressed in fabrication of small-pixel FPAs concern:

- Pixel delineation,
- Pixel hybridization,
- Dark current, and
- Unit cell capacity.

TABLE 4.3    Required $f/\#$ to = 2.0. Real Optics Usually Has $f/\# > 1$ (after Ref. [49])

| $d$ (μm) | MWIR (4-μm) | LWIR (10-μm) |
|---|---|---|
| 2.0 | 1.0 | – |
| 2.5 | 1.25 | – |
| 3.0 | 1.33 | – |
| 5.0 | 2.5 | 1.0 |
| 6.0 | 3.0 | 1.3 |
| 12 | 6.0 | 2.4 |
| 15 | 7.5 | 3.0 |
| 17 | 8.5 | 3.4 |
| 20 | 10.0 | 4.5 |
| 25 | 12.5 | 5.0 |

The above topics are considered in more detail in [64].

Unlike visible sensors, where the pixel size has been reduced to about 1-μm, the scaling of infrared pixels is much more difficult. The detector pixels are generally connected to the per-pixel electronics through indium bump bonds. As the pixel size is reduced, "bump-bonding", ROIC, signal-integrating capacitor, and signal-to-noise ratio all become more difficult.

While bump-bond pitches of 12-μm or larger are relatively common, pitches of less than 8-μm offer significant challenges in terms of array yield, pixel operability, and cost. It is expected that developments will

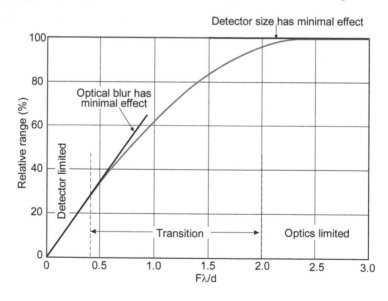

FIGURE 4.29    Relative range as a function of $F\lambda/d$.

continue to extend bump-bonding technology to smaller pixels, as well as to improve manufacturability, and to reduce cost at all pixel sizes. In the past, much of the processing done on a focal plane was confined to the two-dimensional (2D) real estate directly under a given pixel. One novel research area has been the development of 3D integration technologies, providing alternatives to bump bonding [65]. Recently as many as three layers of CMOS have been stacked and vertically interconnected, offering the potential to increase the amount of processing available within a pixel footprint.

In conventional analog ROIC technology, the photocurrent generated by a detector is accumulated and stored locally in a capacitor (electron well); the maximum charge stored during an integration time is equal to the product of the total capacitance and the maximum allowable voltage across the capacitor. A simplified ROIC unit cell (pixel) circuit diagram and a typical ROIC layout are presented in Fig. 4.30. As indicated in this figure, the integrating capacitor dominates the unit-cell area usage. Thanks to multiple gain selection at the pixel level, the charge capacity can be tuned to match the targeted applications and scenario with maximum signal-to-noise ratio (SNR) and dynamic range. The gain and integration time are set, depending on the application requirements ($f/\#$, motion blur,

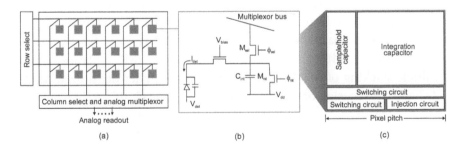

FIGURE 4.30 (a) Analog ROIC architecture (b) with simplified unit-cell or pixel circuit diagram and (c) unit-cell layout. As shown in (b), photocurrent $I_{det}$ is generated by the photodiode and subsequently integrated onto the capacitance $C_{int}$ through the injection transistor $M_i$, which also provides the photodiode bias $V_{bias}$. The signal voltage across $C_{int}$ can be switched onto the multiplexor bus for readout *via* control signal $\phi_{sel}$ on switch $M_{sel}$; the signal voltage across $C_{int}$ can be reset by control signal $\phi_{rst}$ using switch $M_{rst}$. A maximum 2.2-volt process (set by $V_{dd}$) and $C_{int}$ equal to 1850 femtofarads results in a maximum stored photocharge equal to 25 million electrons. Note that circuit capacitance as shown in (c) dominates the pixel footprint (adapted after Ref. [66]).

IR radiation flux), mainly concerning the detection range as well as the detection performance under various weather conditions.

With small pixel pitch, the photon flux is reduced and the integration time is generally increased. As a consequence, the long-range devices involving lower instantaneous field-of-view (IFOV) are often limited by the motion blur, a consequence of holder movement during the integration time. In this case, the impact on range is kept low with a short integration time (2–5 ms). Short integration times are also very useful to freeze a scene with rapidly moving objects. On the other hand, integration time could be increased (10 ms) with a stabilized system independent of the platform's maneuvers and vibrations.

To achieve high sensitivity (e.g. below 30 mK), LWIR FPAs with 5-μm pixels require large amounts of integrated charge to be accommodated in very small unit cells. State-of-the-art storage density of conventional ROIC design rules are about $2.5 \times 10^4$ e⁻/μm². It can provide high-sensitivity values (say < 30 mK) for most tactical MWIR and LWIR applications, utilizing pixel dimensions of 12-μm [62]. For a 5-μm planar unit cell, the charge capacity in standard ROIC technology is less than 1 million electrons, whereas 8 to 12 million electrons are required for high sensitivity. Therefore, small pitch IR detectors are not available today.

The challenge of charge storage in small pixels is being addressed by fabricating microelectromechanical systems (MEMS) capacitors suited to a 3D ROIC design. In recent years, Si-foundries have developed the process technology charge-handling capacities for 2 fF/μm² density metal-insulator-metal (MIM) capacitors in a 0.18-μm CMOS platform [66]. Multiple stacking of MIM capacitors results in higher densities of up to ~7 fF/μm². A redesign of the AIM Infrarot-Module GmbH AIM 640 × 512 15-μm pitch ROIC, in 0.18-μm Si-CMOS technology with MIM capacitor technology, causes improvement of *NEDT* values in comparison with standard ROIC (Fig. 4.31) [67].

Lincoln Laboratory has developed a digital pixel FPA with per-pixel, 16-bit full dynamic range, analog-to-digital conversion, and real-time digital image-processing capability [68]. Developed and tested LWIR ROIC MEMS technology overcomes many performance and scaling limitations imposed by conventional ROIC technology. The MEMS capacitor array can be fabricated in a separate 8-inch (200 mm) wafer. This technology yields 20 million electrons in a 5-μm unit cell. This breakthrough will pave the way for small-pitch FPAs to operate with very high sensitivity. Figure 4.32(b) shows a transmission electron micrograph picture of

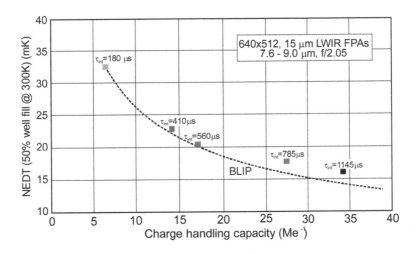

FIGURE 4.31 Measured *NEDT versus* charge-handling capacity for LWIR 640 × 512 HgCdTe 15-μm pitch modules (adapted after Ref. [67]).

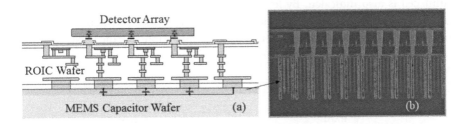

FIGURE 4.32 (a) Schematic illustration of the 3D integrated LWIR FPA design and (b) micrograph of the MEMS capacitor array cross section (adapted after Ref. [68]).

a portion of the MEMS capacitor array. Using the HDVIP technology, a fully functional 1280 × 720, 5-μm unit cell LWIR HgCdTe FPA has been demonstrated.

To maintain or increase the unit cell's dynamic range will require employment of increasingly deeply scaled, higher-density CMOS processes. Higher-density CMOS fabrication processes can also be used to increase in-unit cell processing capacity. The extrapolated unit-cell transistor count, shown in Figure 4.33, suggests the feasibility of advanced in-pixel signal processing in smaller pixels within the next decade. As is predicted in [66], sophisticated in-unit cell processing, coupled with inter-pixel data communication and control structures, would enable massively

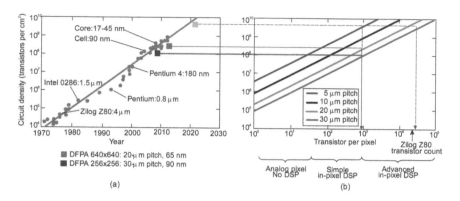

(a)

FIGURE 4.33  (a) The circuit transistor density and trend line for state-of-the-art commercial microprocessors plotted *versus* the year of introduction of each microprocessor to the marketplace. (b) The maximum number of transistors that can be packed into a pixel unit cell as a function of the circuit transistor density for several pixel sizes between 5 and 30 μm. A magnitude estimate of the number of transistors required to achieve three levels of digital processing within the unit cell is also indicated. By leveraging deeply scaled CMOS processes, digital FPA technology enables designers to miniaturize pixel pitch and/or increase on-chip processing capability, depending on application-specific needs (adapted after Ref. [66]).

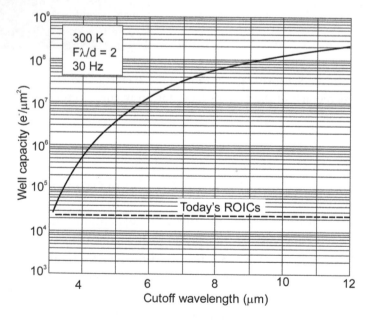

FIGURE 4.34  Required well capacity as a function of cutoff wavelength for BLIP operation at room temperature in the diffraction limit for a 30-Hz frame rate (adapted after Ref. [69]).

parallel computational imagers and resultant sensor systems with capabilities far beyond what is achievable today.

The ultimate desired goal of ROIC for photon detectors is to achieve BLIP operation at room temperature. It requires that the charge from the full $2\pi$ FOV, integrated during a frame time, should be effectively stored on the detector node, even as the pixel pitch is being decreased. Figure 4.34 illustrates the required well capacity for a 5-µm detector pixel as a function of the cutoff wavelength for BLIP operation at room temperature in the diffraction limit. The figure indicates that a significantly larger well capacity per µm² is required to achieve BLIP performance at room temperature, compared with the values available with current ROIC designs.

# REFERENCES

1. W.D. Rogatto (ed.), *The Infrared and Electro-Optical Systems Handbook*, Infrared Information Analysis Center, Ann Arbor and SPIE Press, Bellingham, Washington, 1993.
2. A. Rogalski and F. Sizov, "Terahertz detectors and focal plane arrays", *Opto-Electronics Review* **19**(3), 346–404 (2011).
3. O. Djazovski, "Focal plane arrays for optical payloads", in *Optical Payloads for Space Missions*, chapter 17, ed. Shen-En Qian, Wiley, Chichester, 2016.
4. A. Rogalski, *Infrared and Terahertz Detectors*, 3rd edition, CRC Press, Boca Raton, 2019.
5. P. Norton, "Detector focal plane array technology", in *Encyclopedia of Optical Engineering*, pp. 320–348, ed. R. Driggers, Marcel Dekker Inc., New York, 2003.
6. A. Hoffman, "Semiconductor processing technology improves resolution of infrared arrays", *Laser Focus World*, 81–84 (2006).
7. B. Starr, L. Mears, C. Fulk, J. Getty, E. Beuville, R. Boe, C. Tracy, E. Corrales, S. Kilcoyne, J. Vampola, J. Drab, R. Peralta, C. Doyle, "RVS large format arrays for astronomy", *Proceedings of SPIE* **9915**, 99152X-1-14 (2016).
8. Y. Bai, M. Farris, L. Fischer, J. Maiten, R. Kopp, E. Piquette, J. Ellsworth, A. Yulius, A. Chen, S. Tallarico, E. Hernandez, E. Holland, E. Boehmer, M. Carmody, J.W. Beletic, H. Cho, W. Holmes, M. Seiffert, S. Pravdo, M. Jhabvala, and A. Waczynski, "Manufacturability and performance of 2.3 µm HgCdTe H2RG sensor chip assemblies for Euclid", *Proceedings of SPIE* **10709**, 1070915-115 (2018).
9. A.W. Hoffman, P.L. Love, and J.P. Rosbeck, "Mega-pixel detector arrays: Visible to 28 µm", *Proceedings of SPIE* **5167**, 194–203 (2004).
10. D.A. Scribner, M.R. Kruer, and J.M. Killiany, "Infrared focal plane array technology", *Proceedings of IEEE* **79**, 66–85 (1991).
11. J. Janesick, "Charge coupled CMOS and hybrid detector arrays", *Proceedings of SPIE* **5167**, 1–18 (2003).
12. B. Burke, P. Jorden, and P. Vu, "CCD technology", *Experimental Astronomy* **19**, 69–102 (2005).

13. A. Hoffman, M. Loose, and V. Suntharalingam, "CMOS detector technology", *Experimental Astronomy* **19**, 111–134 (2005).
14. D. Durini (ed.), *High Performance Silicon Imaging*, Elsevier, Cambridge, 2014.
15. W.S. Boyle and G.E. Smith, "Charge-coupled semiconductor devices", *Bell Systems Technical Journal* **49**, 587–593 (1970).
16. J.L. Vampola, "Readout electronics for infrared sensors", in *The Infrared and Electro-Optical Systems Handbook*, Vol. 3, pp. 285–342, ed. W.D. Rogatto, SPIE Press, Bellingham, WA, 1993.
17. E.R. Fossum, "Active pixel sensors: Are CCD's dinosaurs?" *Proceedings of SPIE* **1900**, 2–14 (1993).
18. E.R. Fossum and B. Pain, "Infrared readout electronics for space science sensors: State of the art and future directions", *Proceedings of SPIE* **2020**, 262–285 (1993).
19. M.J. Hewitt, J.L. Vampola, S.H. Black, and C.J. Nielsen, "Infrared readout electronics: A historical perspective", *Proceedings of SPIE* **2226**, 108–119 (1994).
20. L.J. Kozlowski, J. Montroy, K. Vural, and W.E. Kleinhans, "Ultra-low noise infrared focal plane array status", *Proceedings of SPIE* **3436**, 162–171 (1998).
21. L.J. Kozlowski, K. Vural, J. Luo, A. Tomasini, T. Liu, and W.E. Kleinhans, "Low-noise infrared and visible focal plane arrays", *Opto-Electronics Review* **7**, 259–269 (1999).
22. K. Jacobsen, "Recent developments of digital cameras and space imagery", *GIS Ostrava 2011*, 2326 January 2011, Ostrava, http://gisak.vsb.cz/GIS_O strava/GIS_Ova_2011/sbornik/papers/Jacobsen.pdf
23. T. Hirayama, "The evolution of CMOS image sensors", *IEEE Asian Solid-State Circuits Conference*, 5–8, 2013.
24. R. Thorn, "High density infrared detector arrays", U.S. Patent No. 4,039,833, 1977.
25. J.D. Vincent, S.E. Hodges, J. Vampola, M. Stegall, and G. Pierce, *Fundamentals of Infrared and Visible Detector Operation and Testing*, Wiley, New Jersey, 2016.
26. I.M. Baker and R.A. Ballingall, "Photovoltaic CdHgTe-silicon hybrid focal planes", *Proceedings of SPIE* **510**, 121–29 (1984).
27. I.M. Baker, "Photovoltaic IR detectors", in *Narrow-Gap II-VI Compounds for Optoelectronic and Electromagnetic Applications*, pp. 450–473, ed. P. Capper, Chapman & Hall, London, 1997.
28. A. Turner, T. Teherani, J. Ehmke, C. Pettitt, P. Conlon, J. Beck, K. McCormack, L. Colombo, T. Lahutsky, T. Murphy, and R.L. Williams "Producibility of VIP™ scanning focal plane arrays", *Proceedings of SPIE* **2228**, 237–248 (1994).
29. M.A. Kinch, "HDVIP™ FPA technology at DRS", *Proceedings of SPIE* **4369**, 566–578 (2001).
30. *Seeing Photons: Progress and Limits of Visible and Infrared Sensor Arrays*, Committee on Developments in Detector Technologies; National Research Council, 2010, http://www.nap.edu/catalog/12896.html

31. P. Garrou, C. Bower, and P. Ramm (ed.), *Handbook of 3D Integration, Technology and Applications of 3D Integrated Circuits*, 2nd edition, Wiley-VCH, Weinheim, 2008.
32. *Pan-STARRS Releases Largest Digital Sky Survey to the World*, https://www.ifa.hawaii.edu/info/press-releases/panstarrs_release/
33. J.W. Beletic, R. Blank, D. Gulbransen, D. Lee, M. Loose, E.C. Piquette, T. Sprafke, W.E. Tennant, M. Zandian, and J. Zino, "Teledyne Imaging Sensors: Infrared imaging technologies for astronomy & civil space", *Proceedings of SPIE* **7021**, 70210H (2008).
34. G. Konstantatos and E.H. Sargent, "Solution-processed quantum dot photodetectors", *Proceedings of IEEE* **97**(10), 1666–1683 (2009).
35. J.M. Lloyd, *Thermal Imaging Systems*, Plenum Press, New York, 1975.
36. G.C. Holst, "Infrared imaging testing", in *The Infrared & Electro-Optical Systems Handbook*, Vol. 4, Electro-Optical Systems Design, Analysis, and Testing, ed. M.C. Dudzik, SPIE Press, Bellingham, 1993.
37. M.A. Kinch, *State-of-the-Art Infrared Detector Technology*, SPIE Press, Bellingham, 2014.
38. J. M. Lopez-Alonso, "Noise equivalent temperature difference (NETD)", in *Encyclopedia of Optical Engineering*, pp. 1466–1474, ed. R. Driggers, Marcel Dekker Inc., New York, 2003.
39. J.M. Mooney, F.D. Shepherd, W.S. Ewing, and J. Silverman, "Responsivity nonuniformity limited performance of infrared staring cameras", *Optical Engineering* **28**, 1151–1161 (1989).
40. B. Starr, L. Mears, C. Fulk, J. Getty, E. Beuville, R. Boe, C. Tracy, E. Corrales, S. Kilcoyne, J. Vampola, J. Drab, R. Peralta, C. Doyle, "RVS large format arrays for astronomy", *Proceedings of SPIE* **9915**, 99152X-1-14 (2016).
41. J. Johnson, "Analysis of image forming systems", in *Image Intensifier Symposium*, AD 220,160 (Warfare Electrical Engineering Department, U.S. Army Research and Development Laboratories, Ft. Belvoir, pp. 244–273, 1958.
42. U. Adomeit,"Infrared detection, recognition and identification of handheld objects", *Proceedings of SPIE* **8541**, 85410O-1-9 (2012).
43. J.L. Miller, "Future sensor system needs for staring arrays", *Infrared Physics and Technology* **54**, 164–169 (2011).
44. A. Rogalski, J. Antoszewski, and L. Faraone, "Third-generation infrared photodetector arrays", *Journal of Applied Physics* **105**, 091101-44 (2009).
45. S. Horn, P. Norton, T. Cincotta, A. Stolz, D. Benson, P. Perconti, and J. Campbell, "Challenges for third-generation cooled imagers", *Proceedings of SPIE* **5074**, 44–51 (2003).
46. G.C. Holst and T.S. Lomheim, *CMOS/CCD Sensors and Camera Systems*, JCD Publishing and SPIE Press, Winter Park, 2007.
47. J. Robinson, M. Kinch, M. Marquis, D. Littlejohn, and K. Jeppson, "Case for small pixels: system perspective and FPA challenge", *Proceedings of SPIE* **9100**, 91000I-1-10 (2014).
48. R.G. Driggers, R. Vollmerhausen, J.P. Reynolds, J. Fanning, and G.C. Holst, "Infrared detector size: how low should you go?", *Optics Engineering* **51**(6), 063202-16 (2012).

49. G.C. Holst and R.G. Driggers, "Small detectors in infrared system design", *Optics Engineering* **51**(9), 096401-110 (2012).

50. R. Bates and K. Kubala, "Direct optimization of LWIR systems for maximized detection range and minimized size and weight", *Proceedings of SPIE* **9100**, 91000M (2014).

51. C. Li, G. Skidmore, C. Howard, E. Clarke, and C.J. Han, "Advancement in 17 micron pixel pitch uncooled focal plane arrays", *Proceedings of SPIE* **7298**, 72980S–1–11 (2009).

52. R.L. Strong, M.A. Kinch, and J.M. Armstrong, "Performance of 12-μm to 15-μm-pitch MWIR and LWIR HgCdTe FPAs at elevated temperatures", *Journal of Electronic Materials* **42**, 3103–3107 (2013).

53. Y. Reibel, N. Pere-Laperne, T. Augey, L. Rubaldo, G. Decaens, M.L. Bourqui, S. Bisotto, O. Gravrand, and G. Destefanis, "Getting small, new 10 μm pixel pitch cooled infrared products", *Proceedings of SPIE* **9070**, 9070–9094 (2014).

54. Y. Reibel, N. Pere-Laperne, L. Rubaldo, T. Augey, G. Decaens, V. Badet, L. Baud, J. Roumegoux, A. Kessler, P. Maillart, N. Ricard, O. Pacaud, and G. Destefanis, "Update on 10 μm pixel pitch MCT-based focal plane array with enhanced functionalities", *Proceedings of SPIE* **9451**, 9451–9482 (2015).

55. R.K. McEven, D. Jeckells, S. Bains, and H. Weller, "Developments in reduced pixel geometries with MOCVD grown MCT arrays", *Proceedings of SPIE* **9451**, 94512D–1-9 (2015).

56. J.M. Armstrong, M.R. Skokan, M.A. Kinch, and J.D. Luttmer, "HDVIP five micron pitch HgCdTe focal plane arrays", *Proceedings of SPIE* **9070**, 9070–9033 (2014).

57. W.E. Tennant, D.J. Gulbransen, A. Roll, M. Carmody, D. Edwall, A. Julius, P. Dreiske, A. Chen, W. McLevige, S. Freeman, D. Lee, D.E. Cooper, and E. Piquette, "Small-pitch HgCdTe photodetectors", *Journal of Electronic Materials* **43**, 3041–3046 (2014).

58. www.leonardodrs.com/commercial-infrared/products/cooled-camera-modules/hexablu/

59. https://www.lynred.com/

60. G. Destefanis, P. Tribolet, M. Vuillermet, and D.B. Lanfrey, "MCT IR detectors in France", *Proceedings of SPIE* **8012**, 801235-1–12 (2011).

61. N. Oda, private communication.

62. D. Lohrmann, R. Littleton, C. Reese, D. Murphy, and J. Vizgaitis, "Uncooled long-wave infrared small pixel focal plane array and system challenges", *Optical Engineering* **52**(6), 061305-16 (2013).

63. M.A. Kinch, "The rationale for ultra-small pitch IR systems", *Proceedings of SPIE* **9070**, 907032 (2014).

64. A. Rogalski, P. Martyniuk, and M. Kopytko, "Challenges of small-pixel infrared detectors: a review", *Reports on Progress in Physics* **79**(4) 046501 (2016).

65. J. Fan and C.S. Tan, *Low Temperature Wafer-Level Metal Thermo-Compression Bonding Technology for 3D Integration*, InTech, Rijeka, 2012.

66. K.I. Schultz, M.W. Kelly, J.J. Baker, M.H. Blackwell, M.G. Brown, C.B. Colonero, C.L. David, B.M. Tyrrell, and J.R. Wey, "Digital-pixel focal plane array technology", *Lincoln Laboratory Journal* **20**(2), 36–51 (2014).
67. R. Breiter, H. Figgemeier, H. Lutz, J. Wendler, S. Rutzinger, and T. Schallenberg, "Improved MCT LWIR modules for demanding imaging applications", *Proceedings of SPIE* **9451**, 945128-111 (2015).
68. N.K. Dhar and R. Dat, "Advanced imaging research and development at DARPA", *Proceedings of SPIE* 8353, 835302 (2012).
69. M.A. Kinch, "The future of infrared; III–Vs or HgCdTe?", *Journal of Electronic Materials* **44**(9), 2969–2976 (2015).

# CHAPTER 5

# Relevant Properties of Graphene and Related 2D Materials

THE DISCOVERY OF GRAPHENE and other two-dimensional (2D) materials has triggered interest in the development of the next generation of optoelectronic devices, creating a new platform for a variety of photonic applications [1,2], including fast photodetectors [3,4], transparent electrodes in displays and photovoltaic modules [5], optical modulators [6], plasmonic devices [7], and ultrafast lasers [8].

2D materials are atomically thin films, originally derived from layered crystals such as graphite, hexagonal boron nitride (h-BN), the family of transition metal dichalcogenides (TMDs, such as $MoS_2$, $WSe_2$, $MoTe_2$, and others), and black phosphorus (bP). The materials have attracted significant attention, especially in the past decade, owing to their unique and distinctive physical and chemical properties:

- quantum confinement in the direction to the 2D plane, which leads to extraordinary electronic and optical properties, that are of benefit for light absorption,

- a weak stack of atomic planes on top of each other, held by van der Waals (vdW) forces, leaves no dangling bonds, which make it easy to construct vertical heterostructures and to integrate 2D materials with silicon chips,

- the atomically thin characteristic enables scaling-down to nanodevices without parasitic capacitance.

In this chapter, we briefly discuss the fundamental properties of 2D crystals. As there are many comprehensive papers focusing on 2D material synthesis, this field will not be reviewed here. Although 2D materials cover a wide range of compounds across the periodic table and cover a very wide range of the electromagnetic spectrum, only a handful of material groups have been explored for traditional optoelectronic applications. As shown in Fig. 5.1, the 2D materials range from graphene and its zero bandgap to small-/mid-bandgap materials, such as phosphorene and the TMDs, and to wide-bandgap hexagonal boron nitride (h-BN), which is used exclusively as a topologically smooth insulator. Table 5.1 lists the significant

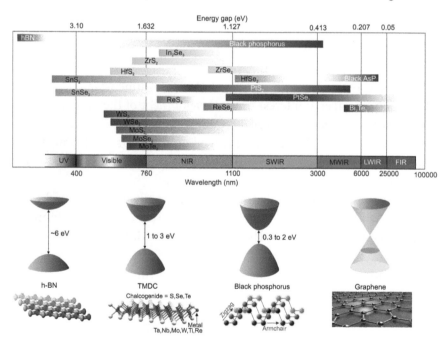

FIGURE 5.1 Bandgap of the different layered semiconductors and the electromagnetic spectrum. The exact bandgap value depends on the number of layers, strain level, and chemical doping. FIR: far infrared; LWIR: long wavelength infrared; MWIR: mid wavelength infrared; SWIR: short wavelength infrared; NIR: near infrared; UV: ultraviolet. The atomic structures of hexagonal boron nitride (h-BN), TMDs, black phosphorus (bP), and graphene are shown at the bottom of the panel, from left to right. The crystalline directions ($x$ and $y$) of anisotropic bP are indicated.

TABLE 5.1 Room Temperature Properties of Selected 2D Materials.

| 2D Material | Bandgap* (eV) | Effective Mass ($m_o$) | Device Mobility ($cm^2/Vs$) | Saturation Velocity (m/s) | Young's Modulus (GPa) | Thermal Conductivity** (W/mK) | CTE*** ($10^6 \, K^{-1}$) |
|---|---|---|---|---|---|---|---|
| Graphene | 0 (D) | <0.01 | $10^3$–$5 \times 10^4$ | (1–5) $\times 10^5$ | 1000 | 600–5000 | – 8 |
| 1L $MoS_2$ | 1.8 (D) | ~0.5 | 10–130 | $4 \times 10^4$ | 270 | 40 | NA |
| Bulk $MoS_2$ | 1.2 (I) | | 30–500 | $3 \times 10^4$ | 240 | 50(‖), 4(⊥) | 1.9(‖) |
| 1L $WSe_2$ | 1.7 (D) | 0.31 | 140–250 | $4 \times 10^4$ | 195 | NA | NA |
| Bulk $WSe_2$ | 1.2 (I) | | 500 | NA | 75–100 | 9.7(‖), 2(⊥) | 11(‖) |
| h-BN | 5.9 (D) | | NA | NA | 220–880 | 250–360 (‖), 2 (⊥) | –2.7 |
| Phosphorene | 0.3–2 (D) | 0.17 | 50–100 | NA | 35–164 | 10–35(‖) | NA |
| bP | 0.3–1.6 (D) | 0.14–0.18 | 500–1000 | ~$10^5$ | ~60 (zigzag) ~27 (armchair) | 60–80 (zigzag) 30 (armchair) | 6–10 |

All listed values should be considered estimates. In some cases, experimental or theoretical values are not available (NA).

*D, I represent direct, indirect energy gap, respectively

**The ‖ symbol signifies the in-plane direction; ⊥ signifies the out-of-plane direction.

***CTE, coefficient of thermal expansion

2D materials used for fabricating optoelectronic devices and a few of their electronic properties.

## 5.1 RELEVANT GRAPHENE PROPERTIES

Graphene has been extensively and comprehensively studied since 2004, due to its unique and exceptional electronic and optical properties [10–12]. The most intriguing and fascinating electronic property of graphene is its linear dispersion relation between the energy and the wave vector where relativistic-like energy dispersion is accompanied by electrons being transported at a Fermi velocity only 100 × lower than the speed of light.

Graphene consists of $sp^2$ hybridized carbon atoms, arranged as a honeycomb, with a lattice constant $a = 1.42$ Å. The valence and conduction bands drop at the Brillouin zone corners (Dirac points), making graphene almost a zero-bandgap semiconductor, as shown in Fig. 5.2. Due to the zero density of states at the Dirac points (the Brillouin zone corners), the conductivity is reasonably low. However, the Fermi level ($E_F$) position can be changed and modified by doping (with electrons or holes) to create a material that is potentially better in terms of conductivity than copper at room temperature. It is commonly known that carbon atoms have a total of six electrons, two in the inner shell and four in the outer shell. The four outer shell electrons in the carbon atom are available for chemical bonding, but, in graphene, each atom is connected to three other carbon atoms on the 2D plane, leaving one electron freely available in the third dimension for electrical conduction. These $\pi$-electrons exhibit high mobility and are located above and below the graphene sheet, where $\pi$-orbitals overlap and enhance the carbon-to-carbon bonds in graphene. Fundamentally,

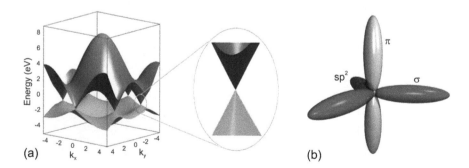

FIGURE 5.2 The band structure of graphene in the honeycomb lattice (a). The enlarged picture shows the energy bands close to one of the Dirac points. Schematic of electron $\sigma$- and $\pi$-orbitals of one carbon atom in graphene (b).

the electronic properties of graphene are determined by the bonding and anti-bonding (the valence and conduction bands) of the π-orbitals.

Graphene exhibits potential for ballistic carrier transport, with the predicted and assessed mean free paths > 2 µm at room temperature, where carriers have been found to spread *via* diffraction, similarly to the light in a waveguide, rather than by carrier diffusion, as happens with a conventional semiconductor.

As mentioned before, the graphene carrier mobility and saturation velocity show potential for use in high-speed photonic devices [13]. A layered graphene structure, with long relaxation times for both electrons and holes, allows for a significant performance improvement in optoelectronic devices. Theoretically, graphene exhibits a room temperature electron mobility of 250000 cm²/Vs; however, the transport mechanism is extremely dependent on the local environment and on material processing. Vacuum-suspended graphene, fabricated by exfoliation, is characterized by extremely high carrier mobilities, >200000 cm²/Vs at room temperature. Unfortunately, these films have been reported to have a very small area (approx. 100 µm²), making them expensive for industrial applications. When placed on a substrate, the graphene mobility is reduced by both charged impurities and remote interfacial phonon scattering effects (Fig. 5.3). On $SiO_2$, interfacial phonon scattering limits graphene mobility to 40000 cm²/Vs [14]. Exposure to atmospheric conditions and processing contaminants, such as resist residue, water, and metallic impurities, also act as scattering sources, limiting mobility.

Another feature which makes graphene interesting in terms of optoelectronic devices is its high thermal conductivity (approx. 10 × copper and 2 × diamond) and high conductivity (approx. 100 × copper). Graphene is also characterized by high tensile strength (130 GPa, compared to 400 MPa for A36 structural steel).

In comparison with the metals, with large quantities of free charges, graphene should be considered to be a semimetal, where carriers can be induced, through chemical doping or electrical gating, with great flexibility, due to their 2D nature, where the doping concentration from $10^{12}$ to $10^{13}$/cm² can easily be reached. Therefore, the semimetal nature of graphene allows for an electrical tunability, which is not feasible for conventional metals.

The optical properties of graphene are also interesting [15]. Graphene's optical conductivity is defined as $\pi\beta$, where $\beta$ is equal to $\left(1/4\pi\epsilon_o\right)\left(e^2/\hbar c\right)$,

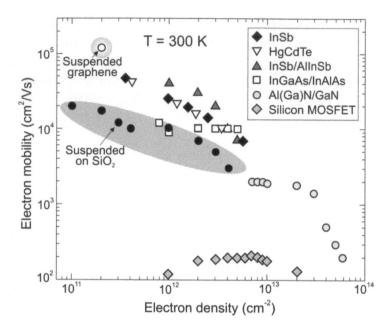

FIGURE 5.3   Electron mobility in graphene at room temperature in comparison with other material systems.

$e$ is the electron charge, $\hbar$ is Planck's constant, $c$ is the speed of light, providing graphene with broadband (visible and infrared, IR) linear absorption of 2.3% per monolayer. The figure of roughly 2.3% of the incident light absorbed by a 0.33-nm graphene monolayer is 10–1000× higher than for semiconductors like silicon and Gallium Arsenide (GaAs), whereas graphene also covers a much broader spectral bandwidth.

Although pristine graphene is a zero-bandgap semiconductor, one can open a bandgap in graphene, using different methods. Graphene's bandgap structure can be modified by substitutional doping [Fig. 5.4(b)], by the addition of two layers [Fig. 5.4(c)] or by bilayer doping [Fig 5.4(d)]. The bandgap of graphene can also be opened by its patterning into a nanoribbon shape or by applying a perpendicular electric field to bilayer graphene. In fact, graphene can cover the range from 0 eV to 0.2 eV. The doping of a graphene layer can move the $E_F$ either up or down, decreasing the mobility of both electrons and holes. Graphene's thickness restriction creates high resistance and chemical inertness, making pure conductive applications less feasible.

Graphene was reported to exhibit the highest-reported specific interaction strength (absorption per atom of material). Silicon has typically a

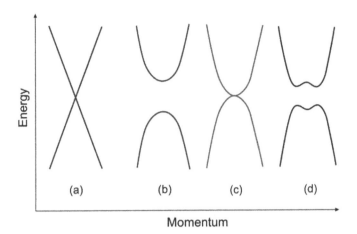

FIGURE 5.4 Modification of the bandgap structure of graphene: the Dirac Fermi cone (a), substitutional doping (b), bilayer graphene (c), and doped bilayers (d).

10-μm absorption depth, causing 2.3% light absorption in a 200-nm layer, whereas graphene reaches the same optical absorption at a much thinner (0.33-nm) layer (interplane spacing).

The absorption spectrum of graphene covers an ultrabroadband range, from visible to terahertz (THz) spectral range [2]. There are two photoexcitation modes: inter-band transition and intra-band transition. Figure 5.5 presents a typical absorption spectrum of doped graphene [16]. For visible and near-infrared (near-IR) light, electrons can be excited from the valence band to the conduction band though the inter-band transition. In the low-frequency THz region, the photon energy is below $2E_F$ and the inter-band transition is prohibited, whereas the intra-band transition dominates. The absorption is due mainly due to the free-carrier (Drude) response. In doped graphene in the mid-IR region, the optical absorption is minimal, and the residual absorption is generally attributed to the disorder in imparting the momentum for the optical transition. A transition occurs close to $2E_F$, where direct inter-band processes lead to a universal 2.3% absorption. In THz band, the coupling of graphene and photons can be enhanced by intra-band transition, and thus it is possible to achieve sensitive THz detection.

## 5.2 PROPERTIES OF 2D CRYSTALLINE MATERIALS

The high dark current of conventional graphene materials, arising from the gapless nature of graphene, significantly reduces the sensitivity of photodetection and restricts further developments of graphene-based

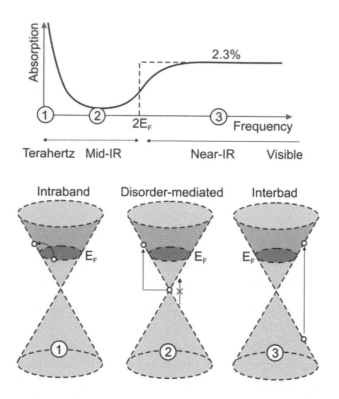

FIGURE 5.5 Characteristic absorption spectrum of doped graphene (adapted after Ref. [16]).

photodetectors. The discovery of new 2D materials, with direct energy gaps from the IR to the visible spectral regions, has opened up a new window for photodetector fabrication. Even though the technology readiness levels are still low, and device manufacturability and reproducibility remain a challenge, 2D material technology can be found in research labs around the globe, including materials like silicene, germanene, stanene, and phosphorene, TMDs, black phosphorus, and the recently discovered all-inorganic perovskites.

The 2D materials are a class of material derived from layered van der Waals (vdW) solids. The in-plane atoms are held together by tight covalent or ionic bonds in 2D directions to form atomic layers, whereas the atomic layers are bonded together by weak vdW interactions along out-of-plane directions. Because of that, a large number of 2D atomic crystals could be mechanically exfoliated from bulk single crystals. Moreover, because of weak physical bonds between the layers, it is possible to mechanically stack arbitrary 2D materials, offering a freedom in fabricating the hetero-structures, based on 2D materials.

Nicolosi *et al.* [17] summarized different types of layered materials, which can be grouped into diverse families Fig. 5.1), which can cover a broad range of electrical and optical properties:

- atomically thin hexagonal boron nitride (h-BN, similar to hexagonal sheets of graphene),

- transition metal dichalcogenides (TMDs),

- black phosphorus (bP), metal halides (e.g. $PbI_2$, $MgBr_2$), metal oxides (such as $MnO_2$ and $MnO_3$), double hydroxides, III–Vs (such as InSe and GaS), V–VIs (such as $Bi_2Te_3$ and $Sb_2Se_3$), and

- halide perovskites.

In this section, we present a short description of the physical properties of 2D materials that offer promising potential for use in next-generation infrared and terahertz detectors. A comparison between the bandgaps of 2D materials is shown in Fig. 5.6, along with those of conventional bulk semiconductors.

## 5.2.1 Transition Metal Dichalcogenides

Layered TMDs are atomically thin materials with the chemical formula of $MX_2$, in which M is a transition metal atom (e.g., W, Mo, or Re) and X is a chalcogen atom (e.g., S, Se, or Te). As is shown at the bottom of Fig. 5.1, one layer of M atoms is sandwiched by two layers of X atoms. The 2D TMDs exist in three polytypes (trigonal, 1T; hexagonal, 2H; and rhombohedral, 3R), which are characterized by different electronic properties, spanning from metallic to semiconducting or even superconducting [18]. The trigonal phase has only been reported in a monolayer shape with a trigonal unit cell, while the 2H and 3R phases have two and three layers, respectively, with hexagonal and rhombohedral unit cells, respectively.

Unlike graphene, where the optoelectronic properties are based on s and p hybridization, the optoelectronic properties of TMDs depend on the d electron count, i.e. filling of the d orbitals of transition metals and its coordination environment. The number of d electron in transition metals varies between zero and six for group 4 to group 10 TMDs, respectively. The completely filled d orbitals, as in the cases of $2H$-$MoS_2$ (group 6) and $1T$-$PtS_2$ (group 10), give rise to a semiconducting nature, while partially filled orbitals, as in the case of $2H$-$NbSe_2$ (group 5) and $1T$-$Re_2$ (group 7), exhibit metallic conductivity.

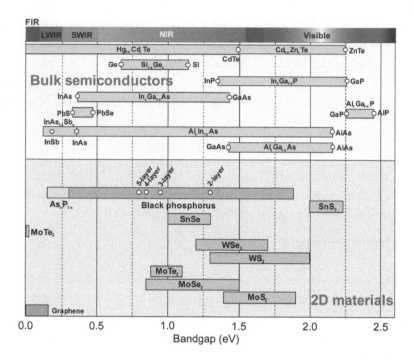

FIGURE 5.6 Comparison of the bandgap values for different 2D semiconductor materials with those of conventional semiconductors. The horizontal bars spanning a range of bandgap values indicate that the bandgap can be tuned over that range by changing the number of layers, straining, or alloying. In conventional semiconductors, the bar indicates that the bandgap can be continuously tuned by alloying the semiconductors (e.g. $In_{1-x}Ga_xAs$ or $Hg_{1-x}Cd_xTe$).

In practical applications, the stability of a material is an important factor, affecting the reliability and lifetime of a device. The electronic properties of TMD materials are mainly determined by the filling of metal atom d orbitals, whereas the lattice parameters and stability primarily depend on the chalcogen atom. The bonding of the M–X is covalent; the metal (M) atom provides four electrons to fill the bonding states, while the lone-pair electrons of the chalcogen (X) atoms terminate the surfaces of the layers. The absence of dangling bonds reduces the chemical instability and protects the surface atoms from reacting with environmental species. Thus, the more stable the lone-pair electrons of the chalcogen (X) atoms are, the more stable the 2D materials will be. This explains, for example, why monolayer $MoS_2$ is more stable than monolayer $MoTe_2$ (Fig. 5.7).

Due to the quantum confinement and surface effects, 2D TMDs exhibit many interesting layer-dependent properties, that differ significantly from

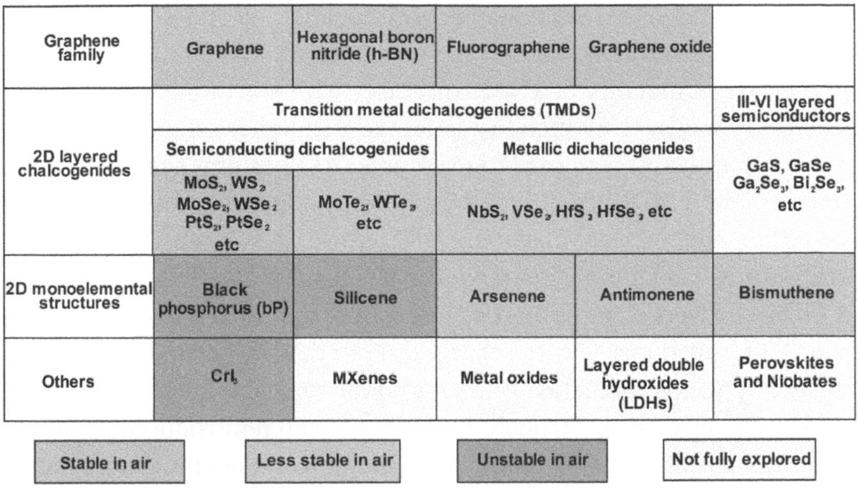

| Graphene family | Graphene | Hexagonal boron nitride (h-BN) | Fluorographene | Graphene oxide | |
|---|---|---|---|---|---|
| 2D layered chalcogenides | Transition metal dichalcogenides (TMDs) | | | | III-VI layered semiconductors |
| | Semiconducting dichalcogenides | | Metallic dichalcogenides | | GaS, GaSe Ga₂Se₃, Bi₂Se₃, etc |
| | MoS₂, WS₂ MoSe₂, WSe₂ PtS₂, PtSe₂ etc | MoTe₂, WTe₂ etc | NbS₂, VSe₂, HfS₂, HfSe₂, etc | | |
| 2D monoelemental structures | Black phosphorus (bP) | Silicene | Arsenene | Antimonene | Bismuthene |
| Others | CrI₃ | MXenes | Metal oxides | Layered double hydroxides (LDHs) | Perovskites and Niobates |

| Stable in air | Less stable in air | Unstable in air | Not fully explored |
|---|---|---|---|

FIGURE 5.7  Summary of stabilities of 2D layered materials (after Ref. [19]).

their bulk crystals. The bulk materials are indirect semiconductors, having a typical bandgap of ~1 eV. In contrast, monolayer TDMs are direct semiconductors with higher bandgaps. As the material becomes thinner from the bulk to the monolayer, the band structure of TMDs transits from a smaller indirect bandgap one to a larger direct bandgap one. In consequence, TMDs can detect light at different wavelengths by tuning the bandgap, by varying the number of layers, due to quantum confinement effects [2,4,20–24]. Moreover, the optical and electronic properties of these materials can be strongly affected by large strains [25,26]. Compared to graphene, TMDs, like molybdenum disulfide (MoS₂), tungsten disulfide (WS₂) and molybdenum diselenide (MoSe₂), exhibit even higher absorption in the visible and the near-infrared (NIR) range and cover a very broad proportion of the spectrum, from the IR to the UV (Figs. 5.1 and 5.6).

The optical absorption of TMDs in visible to NIR regions is dominated by carrier direct transitions between the valence and conduction band states around K and K′ points of the 2D hexagonal Brillouin zone [27], with the contribution from strong excitonic effects. The light absorption can be extended to the mid-IR spectral region because of the existence of defect or edge states inside the bandgap and the relatively high ratios of edge-to-surface area. The absorption coefficient is typically of the order of $10^4$–$10^6$ cm$^{-1}$ so that more than 95% of the sunlight is absorbed by TMD films with sub-micrometer thickness. This high optical absorption can be explained by dipole transitions between localized d states and excitonic coupling of such transitions.

The carrier mobility of TMDs increases with the number of layers; in general, however, their mobility is low (typically less than 250 cm$^2$/Vs) and this disadvantage is hard to overcome. As with graphene, the carrier mobilities of TMDs are limited by ripples, phonon scattering, impurity scattering, and interface scattering [28]. Figure 5.8 summarizes the room-temperature carrier mobility of typical group-6 TMDs, with a comparison of different layers of noble TMDs (PtSe$_2$, PtS$_2$, and PdSe$_2$) and bP on back-gated SiO$_2$ substrates [29]. The charge-carrier density depends on the doping levels and recombination centers, and the typical value is $10^{12}$/cm$^2$ [30].

Very recently, group-10 noble TMDs have been reintroduced as new 2D materials, displaying widely tunable bandgaps, moderate carrier mobility, anisotropy, and ultrahigh air stability. For long-wavelength infrared (LWIR) detector applications, the layered semiconductors, with narrow bandgaps and high mobilities, are required. Among the TMDs, noble metal dichalcogenides provide such opportunities (Fig.re 5.9). It has been theoretically predicted that, at room temperature, the carrier mobility of the group X TMD PtX$_2$ (X = Se, S) is over 1000 cm$^2$/Vs, and the bandgaps of their bilayers and bulks could be very small, at 0–0.25 eV [31,32].

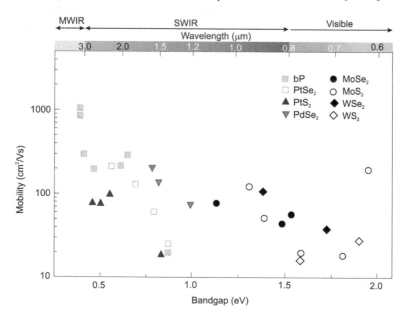

FIGURE 5.8 The layer-dependent room temperature mobility of group-6 TMDCs, bP, and typical noble TMDs on back-gated SiO$_2$ substrate.

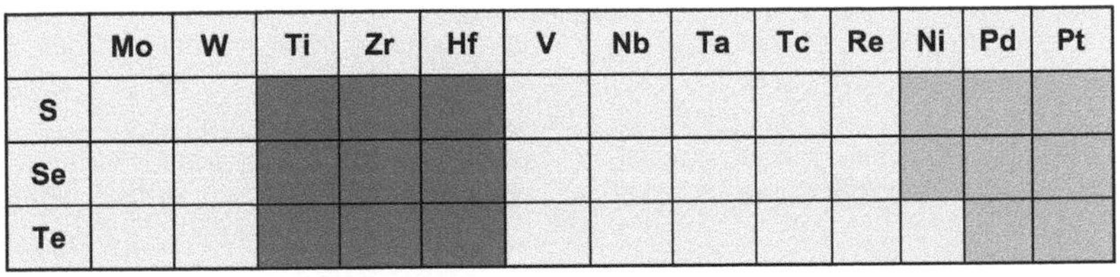

| | Mo | W | Ti | Zr | Hf | V | Nb | Ta | Tc | Re | Ni | Pd | Pt |
|---|---|---|---|---|---|---|---|---|---|---|---|---|---|
| S | | | | | | | | | | | | | |
| Se | | | | | | | | | | | | | |
| Te | | | | | | | | | | | | | |

FIGURE 5.9   Semiconductors among the TMDs with bandgaps below 1 eV (after Ref. [33]).

Unlike most common TMDs, with fewer d electrons, the d orbitals of noble TMDs are nearly fully occupied, and the corresponding $p_z$ orbitals of the interlayer chalcogen atoms are highly hybridized, leading to strong layer-dependent properties and interlayer interactions [29]. The noble metal atoms hold rich d electrons and tend to form $d^2sp^3$ hybridization, where fewer d orbitals are involved, leading to the generation of the thermodynamically favored 1T phase.

The band structure of $MX_2$ (M = Ni, Pt, or Pd; X = S or Se) changes dramatically with the change in layer number. Monolayers $MS_2$ (M = Ni, Pt, Pd) are semiconductors with indirect bandgaps of 0.51, 1.11, and 1.75 eV for $NiS_2$, $PdS_2$, and $PtS_2$, respectively [34]. Furthermore, the $NiS_2$ and $PdS_2$ bilayers become metallic, as predicted by first-principles calculation theory. Experimental results presented in Fig. 5.10(a) verify the layer-dependent bandgaps of $PtS_2$ [29]. As with $PtS_2$, monolayer $PtSe_2$ is also an indirect bandgap semiconductor. First-principles calculation shows that the bandgap of $PtSe_2$ becomes narrow in a bilayer and turns to zero in a trilayer, which basically implies a metallic character [35]. Optical absorption measurements presented in Fig. 5.10(b) confirm that $PtSe_2$ exhibits a gradual transition from a semiconductor (monolayer) to a semimetal (bulk), which is consistent with a theoretical simulation, based on density functional theory. Specifically, when the layer number reaches 50 layers, the bandgap is close to 0.

Despite rapid progress, the technology of the 2D noble TMD family technology is still in its infancy, and there are many challenges to be overcome in this wide-open field.

## 5.2.2 Black Phosphorus

The discovery of black phosphorus (bP), the most stable allotrope of phosphorus, can be dated back a century. It was first synthesized from red phosphorus under high temperature and pressure [37]. Studies of bP as

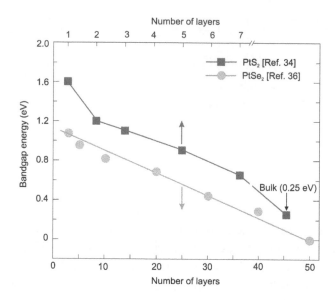

FIGURE 5.10    Layer-dependent bandgaps of noble TMDs: $PtS_2$ and $PtSe_2$.

a bulk material did not receive much attention from the semiconductor research community until 2014. At the beginning of 2014, several research teams reintroduced interest in bP, from the perspective of a layered thin-film material [38,39].

Bulk bP has an orthorhombic structure with a $D_{2h}^{18}$ space group symmetry [22]. In the atomic layer, each phosphorus atom connects to three neighboring atoms, leading to two special directions, namely armchair and zigzag directions, along the $x$ and $y$ axes, respectively (bottom of Fig. 5.1). This highly anisotropic arrangement of phosphorus atoms leads to anisotropic electric band dispersion, further bringing the anisotropic optoelectronic properties. The effective mass of carriers of bP along the zigzag direction is about 10 times larger than that along the armchair direction [40], inducing strong in-plane anisotropy in its electronic, optical, and phonon properties. The strong anisotropic properties can be used to develop new electronic and optoelectronic device applications, such as plasmonic devices, with intrinsic anisotropy in their resonance properties, and high-efficiency thermoelectric, devices exploiting the orthogonality in the heat or electron transport directions [39].

The strong in-plane anisotropy results in a high hole mobility of 1000 $cm^2/Vs$ along the light effective mass direction and about 500 $cm^2/Vs$ along the heavy effective mass direction. At the same time, bP exhibits considerable conductivity in samples with a 25 μm thickness.

The second feature of bP is its wide thickness-dependent bandgaps, which come from the relatively strong interlayer interaction of buckled bP atomic sheets. The bandgap of bP varies with the number of layers (see Fig. 5.11), as has been demonstrated both in theory [41] and by experiment [42], as shown in Fig. 5.11. Obviously, the bandgap of bP increases monotonically as the layer thickness decreases. It should be noted that bP always keeps its direct bandgap nature, despite changes in thickness. It is important to note that the bandgap of bP covers the range 0.3–1.2 eV.

In addition to thickness, the bandgap of bP can also be modulated by other strategies, including applying strain, electric field, or composition alloying. Liu *et al.* [43] have demonstrated full composition tunability of layered $bAs_xP_{1-x}$, covering the long wavelength region down to around 0.15 eV (corresponding to a wavelength of 8.27 μm, in the LWIR regime) (Fig. 5.12). As a result, the bandgap of bP itself (and its compounds) covers an extremely broad range of energy, ~0.15–2 eV, corresponding to ~0.6–8 μm, which has not been achieved in any other 2D layered materials. Thus, bAsP bridges the gap between graphene (a near-zero bandgap semiconductor) and TMDs (a broad bandgap semiconductor).

The transport properties of bP lie between those of graphene and most of the previously studied TMDs, as shown in Fig. 5.13. This figure demonstrates carrier mobilities *versus* current on/off ratio reported for field-effect transistors, based on typical 2D materials. The on/off ratio represents the

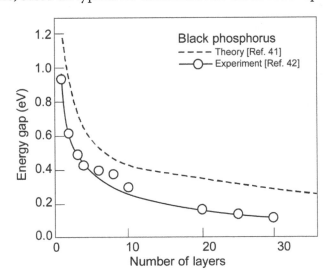

FIGURE 5.11 Thickness-dependent bandgap of bP, both in theory and in experiment.

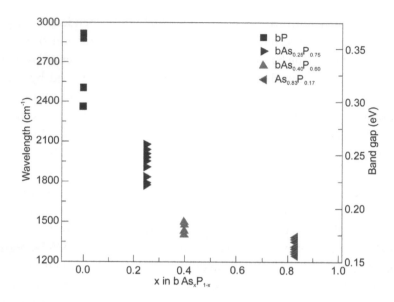

FIGURE 5.12   Bandgaps of b-As$_x$P$_{1-x}$ with various compositions (after Ref. [43]).

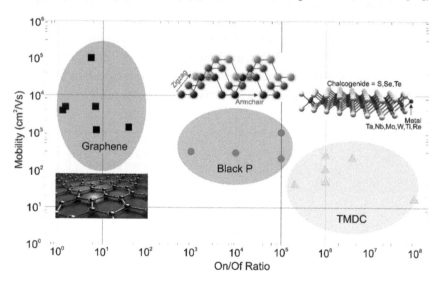

FIGURE 5.13   Carrier mobility *versus* current on/off ratio reported for typical 2D-material electronics (after Ref. [39]).

ratio between the channel currents when the transistor is in the conduction mode ($I_{on}$) and when it is switched off ($I_{off}$). Since the ideal $I_{off}$ value is minimal (to avoid power consumption when not operating), $I_{on}/I_{off}$ should be the highest possible. Typical acceptable values for the on/off ratio are above $10^5$–$10^6$. The onset voltage is defined as the gate voltage necessary for the transistor to switch from the "off state" to the "on state".

Despite the possible variation in mobility in different 2D material classes, these materials fall into three classes. Graphene, with very high mobility, is characterized by the on-off transistor ratio often being less than 10 due to its zero bandgap (high dark current). TMD materials are predisposed toward ultralow power nanoelectronics, whereas black phosphorus falls into the region on the mobility/on-off ratio plot which is not easily covered by either graphene or TMDs. This region, where the mobility is in a range of few hundred $cm^2/Vs$ and the on/off ratio, at the same time, is around $10^4$, is attractive for gigahertz thin-film electronics.

The stability of 2D materials under ambient conditions remains a crucial issue for nanodevices (Fig. 5.7). The absence of high stability in such materials under ambient air conditions has greatly restricted their practical applications. With the rising research interest in bP, studies into the chemical stability of this material are most intensive, owing to its high reactivity and environmental instability under ambient conditions [44,45]. Exfoliated flakes of bP are highly hygroscopic, tending to take up moisture from the air. Long-term contact with the water condensed on the surface degrades the bP. Many researchers have focused on methods to improve its stability in air, using materials and chemicals like $Al_2O_3$, $TiO_2$, $HfO_2$, and titanium sulfonate ligand (TiL4), and coating materials, such as graphene, $MoS_2$, or h-BN [19].

## REFERENCES

1. F. Bonaccorso, Z. Sun, T. Hasan, A.C. Ferrari, "Graphene photonics and optoelectronics", *Nature Photon* **4**, 611–622 (2010).
2. X. Li, L. Tao, Z. Chen, H. Fang, X. Li, X. Wang, J.-B. Xu, and H. Zhu, "Graphene and related two-dimensional materials: Structure-property relationships for electronics and optoelectronics", *Applied Physics Reviews* **4**, 021306-1–31 (2017).
3. F. Xia, T. Mueller, Y.M. Lin, A. Valdes-Garcia, and P. Avouris, "Ultrafast graphene photodetector", *Nature Nanotechnology* **4**, 839–843 (2009).
4. G. Wang, Y. Zhang, C. You, B. Liu, Y. Yang, H. Li, A. Cui, D. Liu, and H. Yan, "Two dimensional materials based photodetectors", *Infrared Physics and Technology* **88**, 149–173 (2018).
5. S. Bae, H. Kim, Y. Lee, X. Xu, J.-S. Park, Y. Zheng, J. Balakrishnan, T. Lei, H.R. Kim, Y.I. Song, Y.-J. Kim, K.S. Kim, B. Özyilmaz, J.-H. Ahn, B.H. Hong, and S. Iijima, "Roll-to-roll production of 30-inch graphene films for transparent electrodes", *Nature Nanotechnology* **4**, 574–578 (2010).
6. M. Liu, X. Yin, E. Ulin-Avila, B. Geng, T. Zentgraf, L. Ju, F. Wang, and X. Zhang, "A graphene-based broadband optical modulator", *Nature* **474**, 64–67 (2011).

7. A.N. Grigorenko, M. Polini, and K.S. Novoselov, "Graphene plasmonics", *Nature Photon* **6**, 749–758 (2012).

8. Z. Sun, T. Hasan, F. Torrisi, D. Popa, G. Privitera, F. Wang, F. Bonaccorso, D.M. Basko, and A.C. Ferrari, "Graphene mode-locked ultrafast laser", *ACS Nano* **4**, 803–810 (2010).

9. D. Akinwande, N. Petrone, and J. Hone, "Two-dimensional flexible nano-electronics", *Nature Communications* **5**, 5678 (2014).

10. K.S. Novoselov, A.K. Geim, S.V. Morozov, D. Jiang, Y. Zhang, S.V. Dubonos, I.V. Grigorieva, and A.A. Firsov, "Electric field effect in atomically thin carbon films", *Science* **306**, 666–669 (2004).

11. K.S. Novoselov, A.K. Geim, S.V. Morozov, D. Jiang, M.I. Katsnelson, I.V. Grigorieva, S.V. Dubonos, and A.A. Firsov, "Two-dimensional gas of massless Dirac fermions in graphene", *Nature* **438**, 197–200 (2005).

12. A.K. Geim and K.S. Novoselov, "The rise of graphene", *Nature Materials* **6**, 183–191 (2007).

13. F. Xia, H. Yan, and P. Avouris, "The interaction of light and graphene: Basic, devices, and applications", *Proceedings of IEEE* **101**(7), 1717–1731, 2013.

14. J.-H. Chen, C. Jang, S. Xiao, M. Ishigami, and M.S. Fuhrer, "Intrinsic and extrinsic performance limits of graphene devices on $SiO_2$", *Nature Nanotechnology* **3**, 206–209 (2008).

15. R.R. Nair, P. Blake, A.N. Grigorenko, K.S. Novoselov, T.J. Booth, T. Stauber, N.M.R. Peres, and A.K. Geim, "Fine structure constant defines visual transparency of graphene", *Science* **320**, 1308 (2008).

16. T. Low and P. Avouris, "Graphene plasmonic for terahertz to mid-infrared applications", *ACS Nano* **8**(2), 1086–1001, 2014.

17. V. Nicolosi, M. Chhowalla, M.G. Kanatzidis, M.S. Strano, and J.N. Coleman, "Liquid exfoliation of layered materials", *Science* **340**, 1226419 (2013).

18. J. Wang, H. Fang, X. Wang, X. Chen, W. Lu, and W. Hu, "Recent progress on localized field enhanced two-dimensional material photodetectors from ultraviolet-visible to infrared", *Small* **13**, 1700894 (2017).

19. X. Wang, Y. Sun, and K. Liu, "Chemical and structural stability of 2D layered materials", *2D Materials* **6**, 042001 (2019).

20. M. Buscema, J.O. Island, D.J. Groenendijk, S.I. Blanter, G.A. Steele, H.S.J. van der Zant, and A. Castellanos-Gomez, "Photocurrent generation with two-dimensional van der Waals semiconductor", *Chemical Society Reviews* **44**, 3691–3718, 2015.

21. M. Long, P. Wang, H. Fang, and W. Hu, "Progress, challenges, and opportunities for 2D material based photodetectors", *Advanced Functional Materials* **29**, 1803807 (2018).

22. F. Wang, Z. Wang, L. Yin, R. Cheng, J. Wang, Y. Wen, T.A. Shifa, F. Wang, Y. Zhang, X. Zhan, and J. He, "2D library beyond graphene and transition metal dichalcogenides: a focus on photodetection", *Chemical Society Reviews* **47**(16), 6296–6341 (2018).

23. T. Liu, L. Tong, X. Huang, and L. Ye, "Room-temperature infrared photodetectors with hybrid structure based on two-dimensional materials", *Chinese Physics B* **28**(1), 017302-1–19 (2019).

24. J. Cheng, C. Wang, X. Zou, and L. Liao, "Recent advances in optoelectronic devices based on 2D materials and their heterostructures", *Advanced Optical Materials* **7**, 1800441 (2019).

25. L. Britnell, R.M. Ribeiro, A. Eckmann, R. Jalil, B.D. Belle, A. Mishchenko, Y.-J. Kim, R.V. Gorbachev, T. Georgiou, S.V. Morozov, A.N. Grigorenko, A.K. Geim, C. Casiraghi, A.H. Castro Neto, and K.S. Novoselov, "Strong light-matter interactions in heterostructures of atomically thin films", *Science* **340**, 1311–1313 (2013).

26. Z. Dai, L. Liu, and Z. Zhang, "Strain engineering of 2D materials: Issues and opportunities at the interface", *Advanced Materials* **31**, 1805417 (2019).

27. K.F. Mark and J. Shan,"Photonics and optoelectronics of 2D semiconductor transition metal dichalcogenides", *Nature Photonics* **10**, 216–226 (2016).

28. S. Manzeli, D. Ovchinnikov, D. Pasquier, O.V. Yazyev, and A. Kis, "2D transition metal dichalcogenides", *Nature Reviews Materials* **2**(8), 1–15 (2017).

29. L. Pi, L. Li, K. Liu, Q. Zhang, H. Li, and T. Zhai, "Recent progress on 2D noble-transition-metal dichalcogenides", *Advanced Functional Materials* **29**, 1904932 (2019).

30. Z. Yang, J. Dou and M. Wang, "Graphene, transition metal dichalcogenides, and perovskite photodetectors", in *Two-dimensional Materials for Photodetector*, ed. Pramoda Kumar Nayak, IntechOpen, 2018. doi: 10.5772/intechopen.74021

31. W. Zhang, Z. Huang, W. Zhang, and Y. Li, "Two-dimensional semiconductors with possible high room temperature mobility", *Nano Research* **7**, 1731–1737 (2014).

32. Y. Zhao, J. Qiao, Z. Yu, P. Yu, K. Xu, S.P. Lau, W. Zhou, Z. Liu, X. Wang, W. Ji, and Y. Chai, "High-electron-mobility and air-stable 2D layered $PtSe_2$ FETs", *Advanced Materials* **29**, 1604230 (2017).

33. P. Wang, Q. Bao, and W. Hu, "Infrared photodetectors", in *2D Materials for Photonic and Optoelectronic Applications*, pp. 105–115, ed. Q. Bao H.Y. Hoh, Elsevier, Amsterdam, 2020.

34. M.R. Habib, W. Chen, W.-Y. Yin, H. Su, and M. Xu, "Simulation of transition metal dichalcogenides", in *Two Dimensional Transition Metal Dichalcogenides. Synthesis, Properties, and Applications*, pp. 135–172, ed. N.S. Arul and V.D. Nithya, Springer, Singapore, 2019.

35. X. Yu, P. Yu, D. Wu, B. Singh, Q. Zeng, H. Lin, Wu Zhou, J. Lin, K. Suenaga, Z. Liu, Q.J. Wang, "Atomically thin noble metal dichalcogenide: a broadband mid-infrared semiconductor", *Nature Communications* **9**, 1545 (2018).

36. L.-H. Zeng, D. Wu, S.-H. Lin, C. Xie, H.-Yu Yuan, W. Lu, S.P. Lau, Y. Chai, L.-B. Luo, Z.-J. Li, Y.H. Tsang, "Controlled synthesis of 2D palladium diselenide for sensitive photodetector applications", *Advanced Functional Materials* **29**, 1806878 (2019).

37. P.W. Bridgman, "Two new modifications of phosphorus", *Journal of the American Chemical Society* **36**(7), 1344–1363 (1914).

38. L. Li Y. Yu, G.J. Ye, Q. Ge, X. Ou, H. Wu, D. Feng, X.H. Chen and Y. Zhang, "Black phosphorus field-effect transistors", *Nature Nanotechnology* **9**(5), 372–377 (2014).

39. X. Ling, H. Wang, S. Huang, F. Xia, and M.S. Dresselhaus, "The renaissance of black phosphorus", *PNAS* **112**(15), 4523–4530 (2015).
40. Y. Akahama, S. Endo, and S. Narita, "Electrical properties of black phosphorus single crystals", *Journal of the Physical Society of Japan* **52**(6), 2148–2155 (1983).
41. V. Tran, R. Soklaski, Y. Liang, and L.Yang, "Layer-controlled band gap and anisotropic excitons in few-layer black phosphorus", *Physical Review B* **89**, 235319 (2014).
42. S. Das, W. Zhang, M. Demarteau, A. Hoffmann, M. Dubey, and A, Roelofs, "Tunable transport gap in phosphorene", *Nano Letters* **14**, 5733–5739 (2014).
43. B. Liu, M. Köpf, A.N. Abbas, X. Wang, Q. Guo, Y. Jia, F. Xia, R. Weihrich, F. Bachhuber, F. Pielnhofer, H. Wang, R. Dhall, S.B. Cronin, M. Ge, X. Fang, T. Nilges, and C. Zhou, "Black arsenic–phosphorus: layered anisotropic infrared semiconductors with highly tunable compositions and properties", *Advanced Materials* **27**, 4423–4429 (2015).
44. J.D. Wood, S.A. Wells, D. Jariwala, K.S. Chen, E. Cho, V.K, Sangwan, X. Liu, L.J. Lauhon, T.J. Marks, and M.C. Hersam,"Effective passivation of exfoliated black phosphorus transistors against ambient degradation", *Nano Letters* **14**, 6964–6970 (2014).
45. J.O. Island, G.A. Steele, H.S.J. van der Zant, and A. Castellanos-Gomez, "Environmental instability of few-layer black phosphorus", *2D Materials* **2**, 011002 (2015).

# CHAPTER 6

# Graphene-Based Detectors

THIS CHAPTER GIVES AN overview of the performance of emerging graphene-based detectors, comparing them with traditional and commercially available ones in different applications under high-operating-temperature conditions. Generally, 2D-material detectors can be divided into two categories, either photon or thermal detectors. Photon detectors are related to the excitation of free carriers as a result of an optical transition, such as the photoconductive or photovoltaic effects. Thermal detectors operate according to thermal effects, such as the bolometric effect or the photothermoelectric (PTE) effect.

## 6.1 TYPES OF DETECTORS

### 6.1.1 Photoconductors

Schematic operations of the two most popular photodetectors are shown in Fig. 6.1. The photoconductive detector is essentially a radiation-sensitive resistor with two metal contacts. A photon of energy greater than the bandgap energy is absorbed to produce electron-hole (e-h) pairs, thereby changing the electrical conductivity of the material. The e-h pairs produced are separated by the external electric field, generating a photocurrent.

Assuming that the signal photon flux density $\Phi_s(\lambda)$ is incident on the detector area $A = wl$ ($w$: width, $l$: length), the basic expression describing photoconductivity in semiconductors under equilibrium excitation (i.e., steady state) is

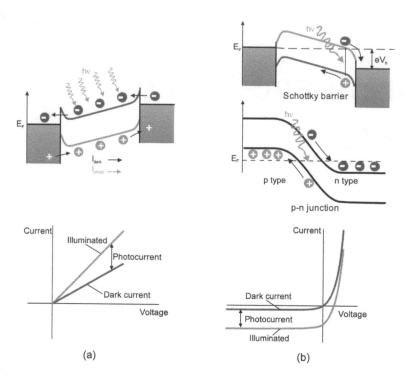

FIGURE 6.1 Schematic of the (a) photoconductive and (b) photovoltaic effects.

$$I_{ph} = q\eta A\Phi_s g,\qquad(6.1)$$

where $q$ is the electron charge, $I_{ph}$ is the short-circuit photocurrent at zero frequency, that is, the increase in current above the dark current accompanying irradiation. The quantum efficiency, $\eta$, can be defined as the number of electron-hole pairs generated per incident photon, and describes how well the detector is coupled to the incoming radiation. The second parameter, the photoconductive gain, $g$, is determined by the properties of the detector (i.e., by which detection effect is used, and the material and configuration of the detector) and can be defined as the number of carriers passing contacts per generated pair. The value of $g$ describes how well the generated charge carriers are used to generate the current response of a photodetector.

In general, photoconductivity is a two-carrier phenomenon and the total photocurrent of electrons and holes is given by:

$$I_{ph} = \frac{qwt\left(\Delta n\mu_e + \Delta p\mu_e\right)V_b}{l},\qquad(6.2)$$

where $\mu_e$ is the electron mobility, $\mu_h$ is the hole mobility; $V_b$ is the bias voltage, and

$$n = n_o + \Delta n; \qquad p = p_o + \Delta p, \qquad (6.3)$$

where $n_o$ and $p_o$ are the average thermal equilibrium carrier densities, and $\Delta n$ and $\Delta p$ are the excess carrier concentrations.

Taking the conductivity to be dominated by electrons (in all known high-sensitivity photoconductors, this is found to be the case) and assuming uniform and complete absorption of the light in the detector, it can be shown [1] that

$$g = \frac{\tau}{l^2 / \mu_e V_b}. \qquad (6.4)$$

So, the photoconductive gain can be defined as

$$g = \frac{\tau}{t_t}, \qquad (6.5)$$

where $t_t$ is the transit time of electrons between ohmic contacts. This means that the photoconductive gain is given by the ratio of free carrier lifetime, $\tau$, to transit time, $t_t$, between the sample electrodes. The photoconductive gain can be less than or greater than unity, depending upon whether the drift length, $L_d = v_d \tau$, is less than or greater than the interelectrode spacing, $l$. The value of $L_d > l$ implies that a free charge carrier swept out at one electrode is immediately replaced by injection of an equivalent free charge carrier at the opposite electrode. Thus, a free charge carrier will continue to circulate until recombination takes place.

Taking into account Eqs. (6.1) and (6.4), the photocurrent

$$I_{ph} = \frac{q \eta A \Phi_s \tau \mu_e V_b}{l^2} \qquad (6.6)$$

is linearly dependent on the photon flux density (i.e., excitation power), the photogenerated carrier lifetime, the electron mobility, and the applied bias.

The current responsivity of the photodetector is equal to

$$R_i = \frac{\lambda \eta}{hc} qg, \qquad (6.7)$$

where $\lambda$ is the wavelength, $h$ is the Planck constant, and $c$ is the velocity of light.

## 6.1.2 Photogating Effect

A particular example of the photoconductive effect is photogating. The photogating effect can be realized in two ways by:

- Generation of e-h pairs, when one type of carrier is trapped by the localized states (nanoparticles and defects), or

- Generation of e-h pairs in trap-states, and one type of carrier is transferred to 2D materials, whereas the other resides at the same place to modulate the layered materials.

In both cases, due to the long carrier lifetime, the enhancement of sensitivity is at the cost of photoresponse speed.

If holes/electrons are trapped in localized states (Fig. 6.2), they act as a local gate, effectively modulating the resistance of active materials. In this case, the photocarriers are limited only by the recombination lifetime of the localized trap states, leading to a large photoconductive gain, $g$. The trap states, where carriers can reside for long periods, are usually located at defects or at the surface of the semiconducting material. This effect is of

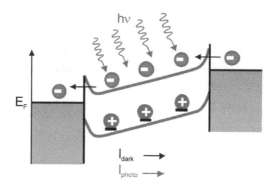

FIGURE 6.2   Band alignment under illumination with photons of energy higher than the bandgap generating e-h pairs. Holes are trapped at the band edge and act as a local gate. In consequence, the field-effect induces more electrons in the channel, generating a photocurrent which adds to the dark current. If the electron lifetime exceeds the time it takes for the electron to transit device, then the long time that the holes are trapped ensures that the electrons can circulate through an external circuit many times, resulting in gain.

particular importance for nanostructured materials, like colloidal quantum dots, nanowires, and 2D-semiconductors, where the large surface and reduced screening play a major role in the electrical properties.

In the case of the photodiode, the photoelectric effect is usually equal to 1, due to separation of minority carriers by the electrical field of the depletion region. However, in a hybrid combination of 2D-material photodetectors, photosensitization and carrier transport take place in separately optimized regions: one for efficient light absorption, and the second, to provide fast charge reticulation. In this way, ultrahigh gain up to $10^8$ electrons per photon, and exceptional responsivities for short-wavelength infrared photodetectors have been demonstrated [2,3].

The simple architecture of a hybrid phototransistor, very popular in the design of 2D-material photodetectors, with the fast transfer channel for charge carriers, is shown in Fig. 6.3. Since, e.g., the graphene in these devices is not responsible for light absorption but only for the sensing of charge, absorber choice determines the spectral response. The graphene's large ambipolar mobility (~$10^3$–$10^5$ cm²/Vs) acts as a built-in photogain (i.e., amplifier) mechanism, enhancing the detector response.

2D materials with thickness down to the atomic layer are more susceptible to local electric fields than are conventional bulk materials, and the photogating effect can strongly modulate the channel conductivity by

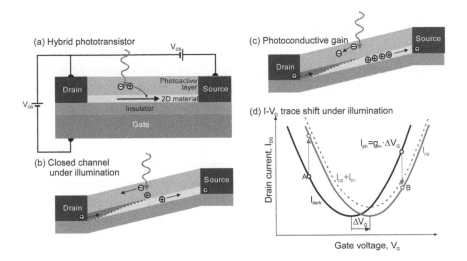

FIGURE 6.3  Photogating effect in 2D-material photodetectors: (a) the operation of hybrid phototransistor, (b) closed channel under illumination, (c) photoconductive gain, and (d) $I$-$V_G$ trace under illumination.

external gate voltage, $V_G$. Improving the optical gain is particularly important since the quantum efficiency is limited because of the weak absorption in 2D materials. This effect is especially seen in longer-wavelength IR spectral regions, where the light absorption is weak. In the case of the hybrid detector shown in Fig. 6.3(a), the holes are injected into the transporting channel, whereas the electrons remain in the photoactive layer. The injected charges can reticulate several thousand times before recombination, giving contribution in photogain under illumination. The photocarrier lifetime is enhanced through both the bandgap structure and defect engineering, and, at the same time, the trapping mechanisms limit the response time of the photodetector, to as much as several seconds. There is a negative trade-off between the enhancement of sensitivity and photoresponse speed.

The photocurrent change by photogating effect can be written as [4,5]:

$$I_{ph} = g_m V_G, \tag{6.8}$$

where $g_m$ is the transconductance and $\Delta V_G$ is the equivalent photoinduced voltage. Figure 6.3(d) indicates a shift of the $I_{DS}(V_G)$ trace after the illumination. Generally, both positive and negative photoconductance behaviors are observed in hybrid 2D structures, and working points $A$ and $B$, related to $g_m$ and $\Delta V_G$, respectively, perform in opposite directions.

### 6.1.3 Photovoltaic Detectors

Photovoltaic (PV) photocurrent generation is based on the separation of electrical carriers by a built-in electric field at a p-n junction or Schottky barrier [top of Fig. 6.1(b)]. By diffusion, the electrons and holes generated within a diffusion length of the junction reach the space-charge region. The generated photocurrent shifts the current-voltage characteristics as is shown in the bottom of Fig. 6.1(b).

The graphene p-n junctions can be feasibly obtained since the Fermi level can be easily tuned because of the limited density of states and ease of formation of p-type or n-type doping. The electrical and optical properties of graphene p-n junctions are different from traditional silicon p-n junctions in terms of the response speed and physical mechanisms involved in photoconversion. Figure 6.4 illustrates the creation of a junction between positively (p-type) and negatively (n-type) doped regions of graphene. The same effect can be reached by applying a source-drain bias, producing an external electric field. Since graphene is a semimetal, generating

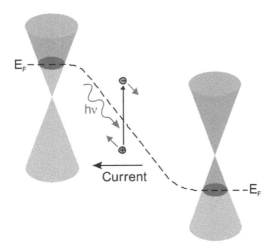

FIGURE 6.4 Schematic of photo-induced extraction of electron-hole pairs and its separation at the graphene p-n junction.

a large dark current, that last approach is usually avoided. The built-in field can be introduced in different ways, either by local chemical doping and electrostatically using split gates, or by exploiting the work-function difference between graphene and the contacts. Typically, p-type doping is reached for metals with a work function higher than that of intrinsic graphene (4.45 eV), whereas the graphene channel can be adjusted to the p- or n- state by the gate.

Photodiodes are usually operated at zero bias (photovoltaic mode) or at reverse bias (photoconductive mode). The absolute response of the photodiode is usually smaller than that of a photodetector working with the photoconducting or photogating mechanisms, since there is no internal gain. Under the reverse-bias operation, the junction capacity is reduced, increasing the speed of the photodiode. Strong reverse bias can initiate impact ionization multiplication of carriers, or avalanching (avalanche photodiode). The large internal gain results in detection of an extremely low signal power. Electron-electron scattering in graphene can lead to the conversion of one high electron-hole pair (e-h) energy into multiple e-h pairs of lower energy, potentially enhancing the photodetection efficiency [6].

Figure 6.5 shows graphene phototransistor architecture, accompanied by the short-circuit photocurrent induced by light. Phototransistors basically have the same three-terminal configuration as field-effect transistors (FETs). In the operational mode of a normal FET, the amount of current

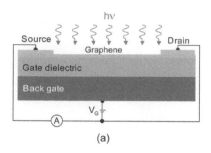
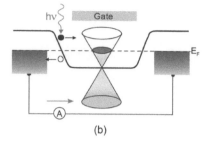

(a)    (b)

FIGURE 6.5   Graphene phototransistor: (a) structure of transistor and (b) a schematic view of photocurrent generation.

flowing (the drain current, $I_d$) in the accumulated channel is controlled by the magnitude of the gate voltage ($V_g$) at a given source-to-drain bias ($V_{ds}$), for phototransistors; the control of channel conductance can additionally be induced by the absorption of light. If there is no bias applied between the source and the drain, minimal photocurrent is collected when the light spot is focused on the middle of the graphene channel. Significant photocurrent is observed when the light is incident on the metal graphene interface area, which is attributed to the conventional PV effect. The built-in electric field in graphene (due to different work functions of the graphene and metal contacts) separates e-h pairs, creating a photocurrent in the external circuit. In the middle of the channel, there is no built-in electric field and, as a result, no photocurrent is observed. The built-in electric field can be further adjusted by a gate bias influencing the photocurrent.

### 6.1.4 Photo-thermoelectric Detectors

The most recent approach (the Seebeck effect) also employs the photo-thermoelectric (PTE) effect to create an electric field, due to electron diffusion into metal contacts. This effect is based on the thermoelectric effect caused by non-uniform light illumination. It can be also created by strong differences in the absorption of distinct parts of the device under uniform illumination. Figure 6.6 illustrates the light-induced heating effect in a semiconductor channel (e.g., in a field-effect transistor), which leads to a temperature gradient.

The internal voltage responsible for current flow is directly proportional to the temperature gradient difference

$$\Delta V = \alpha_S \Delta T, \tag{6.9}$$

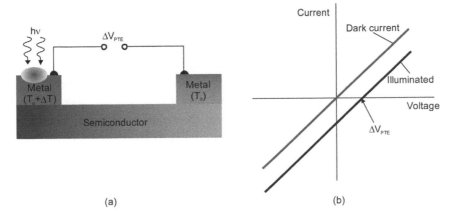

FIGURE 6.6 Photo-thermoelectric effect: (a) in a semiconductor channel and (b) current-voltage characteristics

where $\alpha_S$ is the Seebeck coefficient, commonly expressed in μV/K. The Seebeck coefficient is usually expressed in terms of the conductivity, $\sigma$, of the material [7]

$$\alpha_S = \frac{\pi^2 k^2 T}{3q} \frac{1}{\sigma} \frac{\partial \sigma}{\partial E}, \tag{6.10}$$

where $k$ is the Boltzmann constant, and the derivative of the electrical conductivity, $\sigma$, with respect to energy, $E$, must be evaluated at the Fermi energy [$E = E_F = \hbar v_F k_F$, with $\hbar$ the reduced Planck constant, $v_F$ the Fermi velocity (in graphene about $10^6$ m/s), and $k_F$ the Fermi wavevector]. In the semiconductor, the sign of the Seebeck coefficient is determined by the majority charge polarity.

The coefficient $\alpha_S$ is the effective or relative Seebeck coefficient of the device, composed of two dissimilar conductors "$a$" and "$b$", by electrically joining one set of their ends. Consequently, a thermovoltage is equal to:

$$\Delta V = \alpha_S \Delta T = (\alpha_a - \alpha_b)\Delta T, \tag{6.11}$$

where $\alpha_a$ and $\alpha_b$ are the absolute Seebeck coefficients of the materials, $a$ and $b$. In the example shown in Fig. 6.6(a), with two junctions between the contact metal and the semiconductor, the voltage difference across them is

$$\Delta V_{PTE} = \left(\alpha_{Ssemiconductor} - \alpha_{Smetal}\right)\Delta T \approx \alpha_{Ssemiconductor}\Delta T. \tag{6.12}$$

In the last equation, the term $\alpha_{Smetal}$ is neglected because the Seebeck coefficients of pure metals are in the order 1-$\mu$V/K, much smaller than typical values for semiconductors.

The magnitude of $\Delta V_{PTE}$ is small and typically ranges from tens of $\mu$V to tens of mV (Fig. 6.7). In order to drive current through the device, the ohmic contacts to the semiconductors are required. The weak thermal gradient can be achieved in the case of illumination of a uniform semiconductor, when no current flows in the device, since no external bias is applied. It should be noted that both relative and absolute Seebeck coefficients are temperature dependent and the proportionality between the potential difference generated and the temperature gradient is valid only within the limit of a small temperature difference.

Figure 6.6(b) shows typical current-voltage characteristics of a device with a photoresponse dominated by the photo-thermoelectric effect. Linear dependence of *I-V* curve indicates that the Schottky barriers are small (ohmic contacts). The photothermoelectric effect generates a current at zero bias without changing the resistance.

Here, differences between the photothermoelectric effect and the bolometric effect will be highlighted. In the bolometric effect, the homogeneous temperature change affects the resistivity of a material, and the

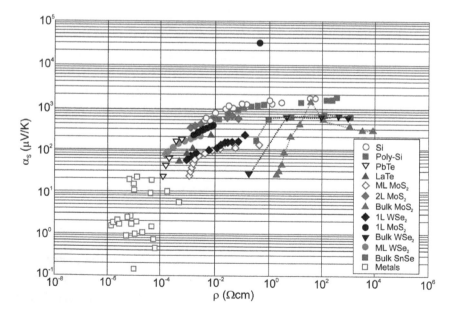

FIGURE 6.7  The Seebeck coefficients as a function of resistivity for various materials (after Ref. [8])

bolometric effect cannot drive a current in a device, only modify the magnitude of the current under external bias and illumination. For a bolometer, the sign of the photocurrent is related to the change in the material conductivity with temperature. In the case of PTE, the sign of the photocurrent is related to the difference in Seebeck coefficients between the components of the junction.

The PTE plays an important role in photocurrent generation [3,9]. For example, in graphene, due to the fact that the optical phonon energy is large (~ 200 meV), hot carriers created by the induced light remain at a temperature higher than that of the lattice for many picoseconds. Equilibrium between the hot electrons and the lattice is reached *via* the slower scattering mechanism between charge carriers (although the charge carriers are substantially speeded-up due to disorder-assisted collisions) and acoustic phonons (nanosecond range). The carrier heat capacity is much smaller than the lattice heat capacity, which leads to a larger temperature gradient in the channel, enhancing the PTE effect. The spot of incident radiation induces carrier temperature variations, and hot carriers generated by photons diffuse due to the temperature gradient, leading to photocurrent generation as was presented in Fig. 6.8, with the carrier and lattice exhibiting different temperatures. The photocurrent direction, due to the PV and PTE effects, is the same, making experimental determination of their relative contributions extremely difficult.

## 6.1.5 Bolometers

The bolometer key parameters are the thermal resistance and the heat capacity. As is shown in Section 3.2, the low- frequency voltage responsivity, $R_v$, of the thermal detector ($\omega \ll 1/\tau_{th}$, where $\tau_{th}$ is the thermal response time) is proportional to the thermal resistance, $R_{th}$, and does not depend on the heat capacitance. At the opposite side, for high frequencies ($\omega \gg 1/\tau_{th}$), the voltage responsivity is not dependent on $R_{th}$ but is inversely proportional to the heat capacitance.

From considerations presented in Chapter 3, the thermal conductance from the detector to the outside world should be small (i.e., the thermal resistance should be high). The smallest possible thermal conductance would occur when the detector is completely isolated from the environment under vacuum, with only radiative heat exchange between it and its heat-sink enclosure.

It was discovered that graphene assumes a low volume for a given area and a low density of states, resulting in a low heat capacity, and exhibiting

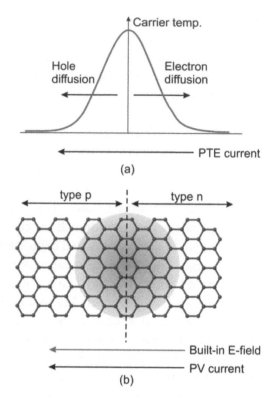

FIGURE 6.8  Photocurrent generation in a graphene p-n junction: (a) profile of carrier concentrations due to light intensity distribution, (b) built-in electric field of p-n junction as well as PTE effect leading to photovoltaic current flow from n-type to p-type region (adapted after Ref [10]).

a fast device response. The electron cooling by acoustic phonons is unproductive (owing to the small Fermi surface) and cooling by optical phonons requires a high temperature ($kT > 200$ meV), leading to the fact that thermal resistance is relatively high, giving rise to bolometric sensitivity [3,11].

Graphene is characterized by over 100-fold anisotropy of heat flow between the in-plane and out-of-plane directions [12]. High in-plane thermal conductivity is due to covalent $sp^2$ bonding between carbon atoms, whereas out-of-plane heat flow is limited by weak van der Waals coupling. The in-plane thermal conductivity of graphene at room temperature is among the highest of any known material, about 2000–4000 W/mK for a freely suspended sample (Fig. 6.9) [12]. The heat flow perpendicular to the graphene sheet is limited with adjacent substrates, e.g. $SiO_2$. It is interesting that the thermal resistance between graphene and its environment dominates that between individual graphene sheets. The discovery

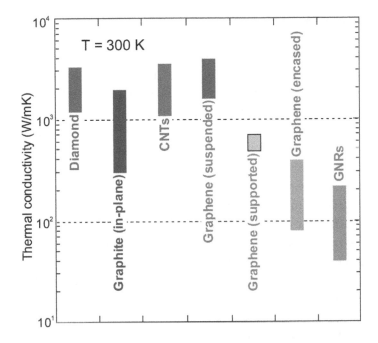

FIGURE 6.9 Room-temperature ranges of thermal conductivity for diamond, graphite, carbon nanotubes (CNTs), graphene, and graphene nanoribbons (GNRs) (after Ref [12]).

of graphene was quickly followed by many other such materials, including $MoS_2$, bP, and hBN, etc.

There are some technical challenges in fabrication of 2D-material bolometers. In graphene, with weak electron-phonon scattering, the resistance is weakly temperature dependent [11]. In consequence, it is challenging to measure the electron temperature change due to incoming radiation power. In addition, to achieve the small electron-phonon thermal conductance, strong thermal isolation is required, which is difficult to achieve. In the case of terahertz detectors, low impedance is required to match the antenna to the external readout circuit, which is also challenging.

2D materials are naturally great candidates for nano electromechanical systems (NEMS) applications. Graphene can sustain large elastic deformations due to its extreme flexibility and could have a positive effect on the device's dynamic range.

### 6.1.6 Field-Effect Transistor Detectors

Nonlinear properties of plasma wave excitations (the electron density waves) in nanoscale field-effect transistor (FET) channels enable their response at frequencies appreciably higher than the device cutoff frequency, which is due to electron ballistic transport. In the ballistic regime of operation, the momentum relaxation time is longer than the electron transit time. The FETs can be used either for resonant (tuned to a certain wavelength) or non-resonant (broadband) THz detection and can be directly tunable by changing the gate voltage.

The graphene FET can also be used for detection of THz radiation, which was first proposed by Dyakonov and Shur in 1993, based on a formal analogy between the equations of the electron transport in a gated 2D transistor channel and those describing the shallow water behavior or acoustic waves in musical instruments, indicating that the hydrodynamic-like effect should exist also in the carrier dynamics in the channel [13]. It must be stressed that instability of that flow in the form of plasma waves was predicted under certain boundary conditions.

The physical mechanism supporting the development of stable oscillations lies in the reflection of plasma waves at the borders of a transistor, with subsequent amplification of the wave's amplitude. Plasma excitations in FETs with sufficiently high electron mobility can be used for emission as well as detection of THz radiation [14,15].

The plasma waves in FETs are characterized by the linear dispersion law [13], and in a gated region:

$$\omega_p = sk = k \left[ \frac{q\left(V_g - V_{th}\right)}{m^*} \right]^{1/2},$$  (6.13)

where $s$ is the plasma wave velocity in the channel, $V_g$ is the gate voltage, $V_{th}$ is the threshold voltage, $k$ is the wave vector, $q$ is the electron charge, and $m^*$ is the electron effective mass.

The plasma wave velocity in the gated region tends to be noticeably larger than the electron drift velocity. A short FET channel, with length $L_g$, acts as a resonant cavity for these waves with the eigenfrequencies $\omega_n = \omega_o(1 + 2n)$ $(n = 1, 2, 3,...)$. The fundamental plasma frequency is

$$\omega_o = \frac{\pi}{2L_g} \left[ \frac{q\left(V_g - V_{th}\right)}{m^*} \right]^{1/2}.$$  (6.14)

When $\omega_o\tau \ll 1$, where $\tau$ is the momentum relaxation time, the detector response is a smooth function of $\omega$ and $V_g$ (broadband detector). When $\omega_o\tau \gg 1$, the FET can operate as a resonant detector, tunable by the gate voltage response frequency, and this device can operate in the THz range. The detection character (resonant or nonresonant) depends on the quality factor of the transistor resonating cavity.

Assuming $m^* \approx 0.1m_o$ ($m_o$ is the free electron mass), $L_g \approx 100$ nm, and $V_g - V_{th} \approx 1$ V, the frequency of plasma waves is estimated to be $\nu_o = \omega_o/2\pi \approx 3$ THz. The minimum gate length can approach $\approx 30$ nm, and, thus, $\nu_o$ can reach 12–14 THz for FETs with GaAs channels.

Summarizing the above discussion, the detection by FETs is due to nonlinear properties of the transistor, leading to the rectification of the AC current induced by the radiation, where photoresponse appears in the form of a DC voltage between the source and the drain, and is proportional to the radiation intensity (PV effect). In the resonant regime, the plasma waves are dimly damped (when a plasma wave launched at the source can reach the drain in a time shorter than the momentum relaxation time) and the detection mechanism exploits interference of the plasma waves in the cavity, resulting in a resonantly enhanced response. Figure 6.10 schematically shows the resonant oscillation of plasma waves in FET's gated region. Even if there is no extra antenna in the system, the THz radiation is coupled to the FET by contact pads and bonding wires. An improvement in sensitivity can be reached by adding a proper antenna or a cavity coupling. Broadband detection occurs when plasma waves are overdamped (meaning that the plasma waves launched at the source decay before they reach the drain).

The photovoltage dependence on carrier density in the FET channel also shows PTE contributions. Figure 6.11 presents and schematically explains two competitive independent detection effects: the plasmonic

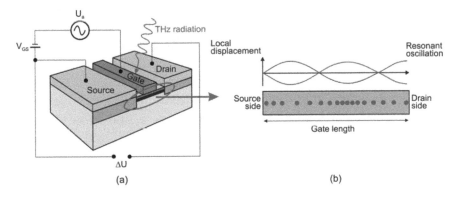

FIGURE 6.10 (a) THz CMOS detector and (b) plasma oscillations in a transistor.

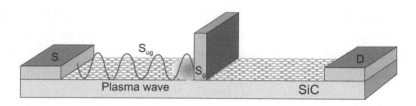

FIGURE 6.11   The detection mechanism in graphene FET THz photodetector.

detection effect due to the nonlinearity of electron transport, and the thermoelectric effect due to the presence of both carrier density junctions and the induced temperature gradient across the FET channel. The red region indicates the locally heated area at the interface of ungated and gated sections with thermopowers $S_{ug}$ and $S_g$, respectively. Even though strongly counterbalanced by the thermoelectric response, the plasma wave detection is the dominant mechanism.

## 6.2  RESPONSIVITY-ENHANCED GRAPHENE-BASED DETECTORS

Many impressive achievements of 2D-based photodetectors have been reported, including ultrafast photoresponse, high responsivity, and ultrabroad detecting band, due to their unique electronic and optoelectronic properties. The majority of pristine graphene photodetectors exploit graphene metal junctions and graphene p-n junctions (to spatially separate and extract the photogenerated carriers), and FET transistors. The development of the graphene high-responsivity photodetectors is determined by two major challenges: the low optical absorption in the detector's active region (~100–200 nm) and the short photocarrier lifetime, meaning that graphene photodetectors are mainly limited by the trade-off between high responsivity, ultrafast time response, and broadband operation.

Responsivity improvement in graphene detectors can be reached by increasing the photocarrier lifetime through both band structure and defect engineering, where carrier trapping mechanisms and patterned graphene nanostructures have been employed to introduce bandgap and mid-gap defect states, whereas the response time is limited by long carrier trapping time due to introduced defect states [3–6,16–19]. Figure 6.12 shows an example of the broadband absorption of a graphene quantum dot (GQD) detector, as reported by Zhang *et al.* [20]. The response was shown from the visible (~ 532 nm), near-IR (~ 1.47 μm) and mid-IR (~ 10 μm) ranges, with photoresponses of 1.25, 0.2, and 0.4 A/W, respectively.

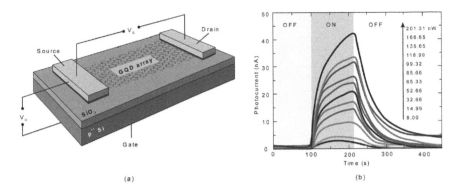

FIGURE 6.12 Broadband GQD photodetector: (a) device design and (b) slow response to light at 1.47 μm (adapted after Ref. [20]).

The single-layer graphene detector's high responsivity [see Fig. 6.12(b)] is partially attributed to internal gain.

Table 6.1 presents several kinds of novel responsivity-enhanced photo-detector structures, consisting of graphene and additional light absorption media (e.g., quantum dots, fluorographene, nanowires, and bulk semiconductors).

The built-in field formed by graphene and light absorption regions can separate the photoinduced carriers generated at the active layer and then inject holes/electrons into the graphene. The photoresponse beyond the light absorption region of semiconductors can also be detected, contributing to the photo-induced carriers provided by the graphene. Unlike the pure graphene photoconductor, the built-in field at the interface can efficiently separate the photo-induced carriers and extend their lifetime (photogating effect), resulting in the relatively high responsivity.

Figure 6.13 shows schematically the differences between a pure graphene photoconductor and a hybrid photoconductor. Additionally, the distinction between ultrafast and ultrasensitive graphene photodetectors is presented. In the early reports, photocurrent was generated by local illumination of one of the metal/graphene interfaces of a back- gated graphene FET. An asymmetric metallization scheme was used to break mirror symmetry of the built-in potential profile within the channel, giving the overall photocurrent, where metal fingers were used, leading to the creation of a significantly enlarged light detection region exhibiting a high electric field [see Fig. 6.13(a)]. The graphene high carrier mobility and short carrier lifetime [see Fig. 6.13(c)] allow metal-graphene-metal photodetectors to operate at high data rates.

TABLE 6.1 Responsivity-Enhanced Graphene-Based Detectors

| | Advantages | Disadvantages | References |
|---|---|---|---|
| Hybrid graphene quantum dots (GQD) detector | Increased absorption and introduction of large carrier multiplication factors | Bandwidth and response time are limited by the narrow spectral bandwidth and long carrier trapping times of the quantum dots | [2,21–25] |
| Graphene-fluorographene detector | Three orders of magnitude- enhanced responsivity compared to pristine graphene detectors. Broadband photoresponse from the ultraviolet to the mid-infrared wavelengths | Fluorographene partially decomposes over time. The slow response time is given by the trap states in the fluorographene. | [26] |
| Two graphene layers separated by a thin tunnel barrier | Broadband responsivity *via* separation of the photogenerated electrons and holes through quantum tunneling and minimization of their recombination | Response times limited by the long carrier trapping times in the tunneling barriers | [27,28] |

(Continued)

TABLE 6.1 (CONTINUED)  Responsivity-Enhanced Graphene-Based Detectors

| | Advantages | Disadvantages | References |
|---|---|---|---|
| Waveguide-integrated graphene detector | Ultrafast responsivity by increasing the interaction length of light within graphene and processing corresponding to standard photonic integrated circuits | Spectral bandwidth restricted by the bandwidth limitations of the waveguides | [29–32] |
| Microcavities, plasmonic structures, and optical antenna integrated with graphene | High responsivities by increasing the interaction length of light within graphene | Bandwidth is limited by the resonant nature of the structures | [9,33–46] |

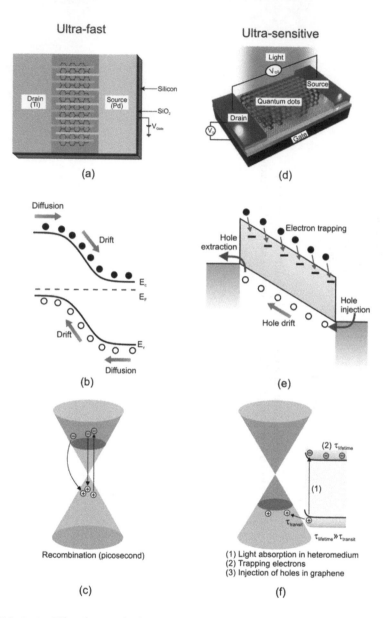

FIGURE 6.13  Ultrafast and ultrasensitive graphene photodetectors: (a) structure of metal-graphene-metal photodetector, (b) band profile, (c) recombination
mechanism, (d) hybrid GQD photodetector (after Ref. [2]), (e) trapping process,
and (f) process dynamics at the interface of graphene/quantum dots.

The main feature of the hybrid photodetector [see Fig. 6.13(d)] is ultra-high gain, originating from the high carrier mobility of the graphene sheet and the recirculation of charge during the lifetime of the carriers, remaining trapped either in the quantum dots [Fig. 6.13(e)] or other light-absorbing regions, to include carbon nanotubes and nanoplates. Photoexcited holes in the quantum dots are transferred to the graphene layer and drift by bias $V_{ds}$ to the drain, with a typical transit time, $\tau_{transit}$ being inversely proportional to the carrier mobility, whereas electrons remain trapped (with a typical lifetime, $\tau_{lifetme}$) in the quantum dots (see Section 6.1.2). Multiple circulation of holes in the graphene channel, following a single e-h photogeneration, leads to a strong photoconductive gain, $g = \tau_{lifetme}/\tau_{transit}$, indicating the importance of long lifetime and high carrier mobility. Konstantatos et al. has demonstrated the gain of $10^8$ electrons per absorbed photon and a responsivity of ~ $10^7$ A/W in short-wavelength (SWIR) hybrid phototransistors [2].

Hybrid photodetectors offer improvements in responsivity, although the majority of these devices have a limited linear dynamic range, due to the charge relaxation time, which quickly saturates the available states for photoexcitation, leading to a drop in responsivity with incident optical power (see Fig. 6.14). Figure 6.14 also shows the gain as a function of excitation intensity, as compared with the theory (solid line), for a hybrid

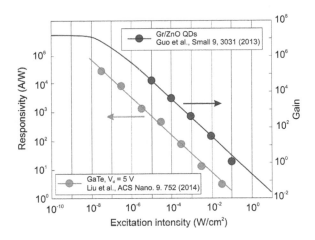

FIGURE 6.14 Responsivity and gain as a function of excitation intensity for for multilayer GaTe flakes [44] and hybrid graphene/ZnO quantum dots [23] detectors, respectively. The circles are experimental data, and the solid curves are the theoretical plots with best fitting.

single-layer graphene/ZnO quantum dots (QDs) detector. This ZnO QDs/graphene hybrid structure shows a photoconductive gain as high as $10^7$.

The light-matter interaction in 2D materials, to improve performance of the photodetectors, can be also achieved by introducing optical structures (e.g., plasmonic nanostructures, photonic crystals, optical cavities, waveguides) on the top of the device. Two key factors are important: size- and shape-matching of the metal pattern to generate plasmons, and the coupling of the plasmons to the detector. The generation of plasmons depends significantly on the metallic pattern. The dimensions of the metal grating should be similar to the width of the metal strip, permitting the plasmon fields to enter the detector below the grating. Usually, a thin dielectric layer is placed over the detector and a metal grid is placed on the top of the dielectric layer. As the plasmons travel parallel to the surface, a large optical path can be reached for absorption, without the need for a thick active layer.

The most advantageous approach to combining high responsivity with fast photoresponse time is to improve the generation rate of photoinduced carriers in the graphene and to maintain the appropriate carrier lifetime. That approach has been demonstrated for SWIR graphene photodetectors [47] (Fig. 6.15). Under illumination, the light with wavelength-matching plasmonic resonance is trapped by Au-nanoparticles and is absorbed by the graphene. A vertical built-in field is employed in the graphene channel

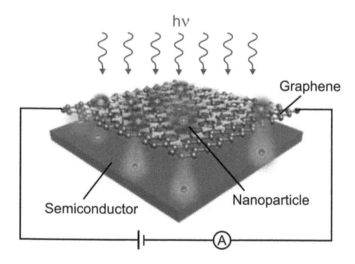

FIGURE 6.15 The concept of SWIR graphene photodetector (adapted after Ref. [47]).

for trapping the photoinduced electrons and leaving holes in the graphene, resulting in prolonged photoinduced carrier lifetime.

The detector's responsivity is enhanced by plasmonic Au nanoparticles and is the highest reported among the SWIR of the graphene-based detectors (83 A/W at 1.55-μm), whereas response time is limited approximately to the level of 600-ns, due to the traps in both the surface of the hybrid structure and in the nanoparticles. In spite of that, the SWIR graphene detector is characterized by the fastest response time of the hybrid graphene photodetector/transistor.

The unique electrical and optical characteristics of gold-patched graphene nanostripe photodetectors have been demonstrated by Cakmakyapan *et al.* [48]. Commercially available CVD-grown graphene was first transferred to a high- resistivity silicon wafer covered with a 130-nm thick thermally grown $SiO_2$ layer. Next, gold patches, graphene nanostripes, Ti/Au contacts, and gate pads were patterned by different combinations of optical lithography and plasma etching. The $V_g$ applied to the Si substrate controls the $E_F$ of the graphene nanostripes. The gold patches have a width of 100-nm, a periodicity of 200-nm, height of 50-nm, length of 1-μm, and a tip-to-tip gap size of 50-nm (Fig. 6.16).

The photodetector has an ultrabroad spectral response from the visible to the IR region, with high-responsivity levels ranging from 0.6 A/W, at a wavelength of 800 nm, to 11.65 A/W at 20 μm, as shown in Fig. 6.17. That wide photodetection bandwidth and high responsivity are achieved by the gold-patched graphene nanostripes. As expected, higher photoconductive gains are obtained at lower wavelengths, where excitation to higher energy

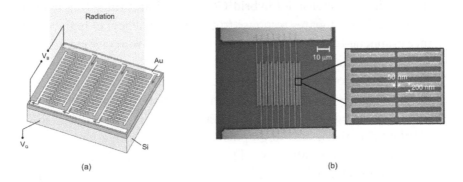

(a)                                    (b)

FIGURE 6.16 Photoconductive nanostructures based on gold-patched graphene nanostripes: (a) the photodetector principle of operation and (b) optical microscope and SEM images for a fabricated photodetector (adapted after Ref. [48]).

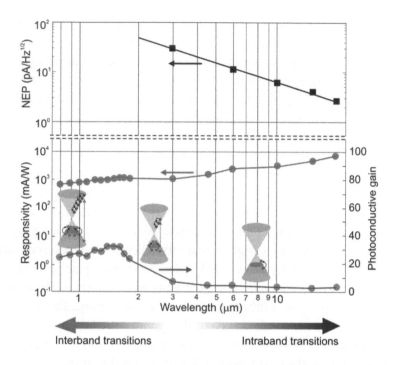

FIGURE 6.17 Responsivity, photoconductive gain, and *NEP* (optical chopping above 1 kHz) of the photodetector at an optical power of 2.5 μW, gate voltage of 22 V, and bias of 20 mV (adapted after Ref. [48]).

levels contributes to the generation of subsequent e-h pairs by energy transfer during relaxation (as shown in the insert of Fig. 6.17).

The photodetector also shows a frequency response exceeding 50 GHz, which is more than seven orders of magnitude faster compared with the higher-responsivity hybrid GQD [2,23] and tunneling barriers [27]. Figure 6.18 compares response (time and frequency) of high-performance graphene-based photodetectors operating at room temperature, as reported in the literature.

Another way to enhance the optoelectronic properties of graphene is by modification by noncovalent and covalent functionalization. Due to its robust chemistry, graphene itself is a chemically inert material. The so-far established chemistries led to graphene derivatives with a low degree of functionalization (typically 1–3%) [21]. Fluorographene (FGr) is prepared by fluorination of graphene, and mechanical or chemical exfoliation of graphite fluoride. The bandgap of FGr can be tuned from the ultraviolet to the near-infrared by controlling the degree of fluorination. In Ref. [26], a Gr/FGr photodetector has a spectral range spanning from 255 nm to 4.3

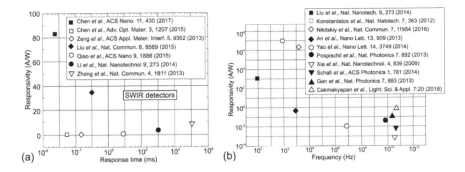

FIGURE 6.18 Comparison of the responsivity and response time and frequency for the room-temperature graphene photodetectors reported in the literature, mainly in (a) Ref. [47] and (b) Ref. [48].

μm ([26], Fig. 3.15). This broadband response arises from the quantum confinement of graphene regions by fluorine adatoms. The rehybridization of carbon with fluorine results in a mixture of $sp^2$ and $sp^3$ nanodomains, inducing a series of discrete states for trapping photoexcited charge carriers. Despite the high photoresponsivity of the Gr/FGr photodetector over a broadband range, its operation speed is slow, about 200 ms. It is suggested that the reason for this is the long trapped carrier lifetime in both the $sp^2$ and the $sp^3$ domains.

## 6.3 GRAPHENE-BASED THERMAL DETECTORS

Thermal detectors are classified, according to the operating schemes, as thermopiles, bolometers, or pyroelectric detectors.

The basic structure of a thermoelectric graphene photodetector design is shown in Fig. 6.19 [8,49]. This device, made of a sheet of graphene with

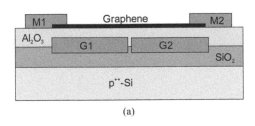

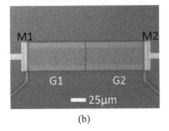

FIGURE 6.19 Schematic graphene split-gate thermopile with (a) supported substrate and (b) its microscopic image. M1 and M2 are metal contacts to graphene, and G1 and G2 are split gates that electrostatically dope the graphene channel to form a p-n junction (after Ref. [8]).

dual split-backgates, develops a photovoltage across electrodes M1–M2 as a function of the voltage applied to the backgates. The device shows a photoresponse, even at 10.6 μm, which indicates that absorption is not limited as a result of Pauli blocking in graphene, but that most of the light absorption is done in the substrate underneath, whereas the graphene devices use the thermoelectric effect to convert the temperature rise in the substrate to a voltage difference, as described by Eq. (6.11): $\Delta V = (\alpha_p - \alpha_n)\Delta T$, where $(\alpha_p - \alpha_n)$ is the difference of the Seebeck coefficient between the p- and n-regions of graphene, and $\Delta T$ is the temperature difference between the graphene p-n junction and the metal contacts.

Hsu *et al.* have demonstrated a graphene-based thermal imaging system by integrating graphene-based photothermo-electric detectors with micromachined silicon nitride membranes [50]. As shown in Fig. 6.19, multiple graphene photodetectors are combined into the thermopile, composed of an infrared absorber, that is suspended from the substrate, and a series of thermal arms that connect the absorber and the surroundings, with interleaved p- and n-type graphene channels on top. Incidence of infrared radiation causes heating of the absorber (dielectric multilayer thin film), which can then be probed electrically by the graphene p-n junctions, due to the thermoelectric effect. Thanks to the use of an optimized IR absorption layer made of a $SiO_2/Si_3N_4/SiO_2$ combination, deposited using plasma-enhanced chemical vapor deposition (PECVD), the absorption in the 8–12 μm spectral range was achieved at a rate > 40%. The fabrication of a free-standing absorber membrane was made after undercutting the silicon underneath with $XeF_2$ isotropic etching.

Figure 6.20 shows multiple graphene photodetectors, combined into the thermopiles, composed of an infrared absorber, that is suspended from the substrate, and a series of thermal arms that connect the absorber and the surroundings, with interleaved p- and n-type graphene channels on top.

It can be shown that, if the thermopile's detectivity is limited by Johnson noise, then $D^{*2}/\tau$ (where $\tau$ is the response time) is independent of the lateral geometry [8]:

$$\frac{D^{*2}}{\tau} = \frac{\alpha_{abs}}{t} \cdot \frac{\Delta \alpha_S^2}{\rho_{2D}} \cdot \frac{1}{k_{th}c_v} \cdot \frac{1}{16kT} \qquad (6.15)$$

Here, the first term $(\alpha_{abs}/t)$ is absorbance per unit thickness, $t$, indicating the capability of IR absorption of the absorber; the second term, $\Delta/\rho_{2D}$, with 2D resistivity of graphene, is determined by the electrical and

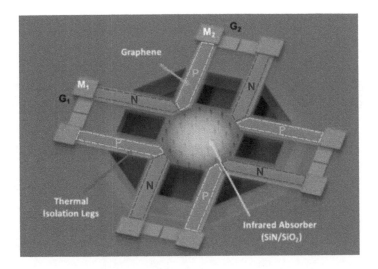

FIGURE 6.20   Graphene thermopile with suspended IR absorber: (a) schematic of a graphene thermopile; the red and blue regions indicate the p-type and n-type region of graphene, and the square in the center is the dielectric absorber. The whole structure is suspended on the substrate to reduce the thermal conductance in the vertical direction (after Ref. [8]).

thermoelectric properties of the sensing material. The third term, $1/\kappa_{th}c_v$, indicates the quality of thermal isolation, with $\kappa_{th}$ denoting the thermal conductivity of the absorber, and $c_v$, the specific heat capacity.

Equation (6.15) contains $\Delta\alpha_S^2/\rho_{2D}$, the thermoelectric figure of merit (FOM) for thermopile IR detectors, and $1/\kappa_{th}c_v$, the thermal transport factor. The Seebeck coefficient *versus* resistivity relationship for different 2D and 3D bulk materials is shown in Fig. 6.7. In order to benchmark graphene-based thermoelectric detectors with respect to the other material systems, the FOMs are plotted as a function of resistivity in Fig. 6.21. As is shown, the FOM for today's standard CVD graphene on $SiO_2$, with an average mobility of 2000 cm²/Vs, can already outperform any thermopiles made with metals and most of the thermoelectric (TE) materials. It is predicted that the use of higher-quality, properly passivated graphene could make the FOM two orders of magnitude higher than that of any of the other material systems [8]. We can also see that the FOM of 2D TMDs is higher than in their 3D counterparts, which indicates their great potential for thermal detection and other thermoelectric applications.

Figure 6.22 compares the detectivity and the response time of graphene thermopiles with different types of state-of-the-art thermal detector

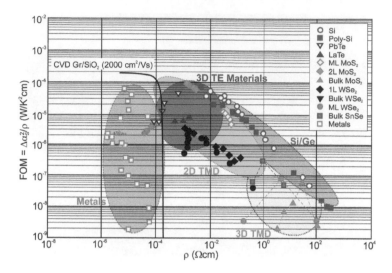

FIGURE 6.21 The thermoelectric figure of merit as a function of resistivity for various materials (after Ref. [8]).

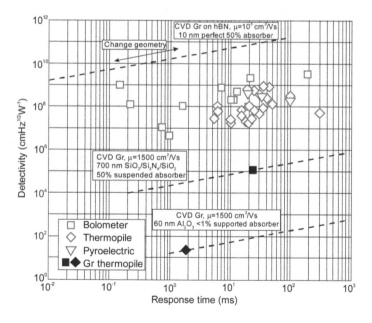

FIGURE 6.22 Detectivity versus response time for different technology nodes of graphene thermopiles in comparison with mainstream uncooled thermal IR detectors (after Ref. [8]).

technologies, including bolometers (VO$x$, etc.), thermopiles (poly-Si, Al, thermoelectric materials, etc.), and pyroelectric devices (PZT and other piezoelectric materials). For more advanced thermal detector technologies, the detectivity magnitude at room temperature is of the order of $\sim 10^8$–$10^9$ cmHz$^{1/2}$/W. As is shown, the performance of current graphene thermopile technology is considerably inferior in comparison with the state-of-the-art thermal detectors, and is below $10^6$ cmHz$^{1/2}$/W. However, the theoretically predicted performance is even better than today's state-of-the-art technologies. For example, a 10-nm thick absorber with good mechanical stability and 50% absorption achieved through nanophotonic structures, would make graphene thermopiles better than any existing bolometers.

The bolometer key parameters are thermal resistance and heat capacity. It has been shown that graphene assumes a low volume for a given area and a low density of states, resulting in a low heat capacity exhibiting a fast device response.

Two types of graphene-based bolometers are shown in Fig. 6.23. Yan *et al.* has considered graphene as a hot-electron bolometer [51], as shown in Fig. 6.23 (a). Due to weak electron-phonon interactions, they used bilayer graphene, exhibiting a tunable bandgap. Implementation of a perpendicular electric field gives rise to electron temperature-dependent resistance at low temperatures, making the device suitable for thermometry. The extrapolated *NEP* value for a 1-µm$^2$ sample at 100 mK is approx.

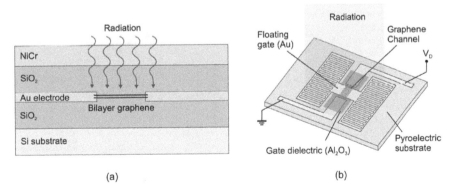

FIGURE 6.23 The graphene bolometers: (a) (a) side view of the bilayer graphene hot-electron bolometer (semitransparent NiCr top gate covers the graphene device and SiO$_2$ surrounds the graphene) and (b) pyroelectric bolometer (conductance of graphene channel is modulated by the pyroelectric substrate and by a floating gate).

$5 \times 10^{-21}$ W/Hz$^{1/2}$, similar to the TES (transition edge sensor) state-of-the-art bolometer. The graphene-based detector structure, with a temperature coefficient of resistance (TCR) above 4%/K, is shown in Fig. 6.23(b), where the pyroelectric response of a LiNbO$_3$ crystal is transduced with high gain (up to 200) into modulation for graphene [52]. That effect is reached by fabrication of a floating metallic structure, concentrating the pyroelectric charge on the top-gate capacitor of the graphene channel.

The main difficulty in development of high-sensitivity graphene bolometers is the weak variation of electrical resistance *versus* temperature. The paper published by El-Fatimy *et al.* [53] has shown that graphene quantum dots on SiC exhibit extremely high variation of resistance *versus* temperature, due to quantum confinement, greater than 430 MΩ/K at 2.5 K, leading to responsivities of $1 \times 10^{10}$ V/W for the THz region. In hot-electron bolometers with quantum dots in epitaxial graphene, the bandgap is induced *via* quantum confinement (without the need for gates), using a simple single-layer structure. Figure 6.24(a) presents the *NEP* for 0.15 THz *versus* temperature from 2.5 K to 80 K, calculated for 30-nm and 150-nm quantum dots. The *NEP* is approximately one order of magnitude lower than the best commercial cooled bolometer and much faster in response time (a few nanoseconds, compared to milliseconds for commercial bolometers). These quantum dot bolometers operate in a very broad spectral range, from THz to ultraviolet radiation, with responsivity being independent with respect to wavelength [Fig. 6.24(b)]. As for hybrid

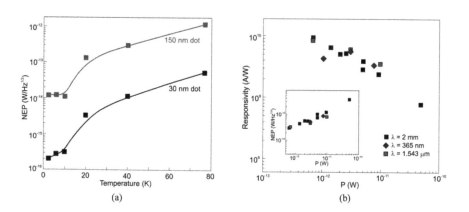

FIGURE 6.24 Quantum dot bolometers: (a) *NEP versus* temperature at 0.15 THz for 30-nm and 15-nm quantum dots (adapted after Ref. [53]) and (b) responsivity *versus* absorbed power at different wavelengths. Inset: *NEP versus* absorbed power at selected wavelengths (adapted after Ref. [54]).

photodetectors (Fig. 6.14), a drop in responsivity *versus* absorber power is also observed for graphene bolometers [Fig. 6.24(b)].

Graphene, with the lowest mass per unit area of any material, extreme thermal stability, and an unmatched spectral absorbance, generates interest as an active bolometer absorber. However, due to its weakly temperature-dependent electrical resistivity, graphene has failed to challenge the state-of-the-art materials at room temperature. Both the speed and sensitivity are inversely proportional to the thermal resistance, so a sensitive bolometer is often slow. A common method to modify the speed and sensitivity is to change the thermal resistance between the bolometer and its environment.

In a departure from conventional bolometer, a graphene nanoelectromechanical system has been proposed to detect light *via* resonant sensing. In the design proposed by Blaikie *et al.* [55], absorbed light heats and thermally tensions a suspended graphene resonator, thereby shifting its resonant frequency. Figure 6.25 illustrates a schematic design of a graphene resonator, comprising a suspended graphene membrane. The graphene nanomechanical bolometer (GNB) is made by transferring graphene onto a Si/SiO$_2$ support substrate with patterned holes, resulting in circular drumhead resonators. Some drumheads are patterned into trampoline geometries, using a focused ion-beam technique. The most sensitive device has a 6-μm diameter trampoline with 200-nm wide tethers.

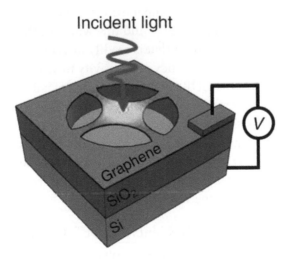

FIGURE 6.25   Design of nanomechanical graphene resonator (after Ref. [55]).

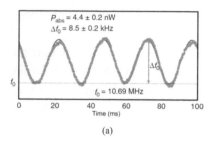

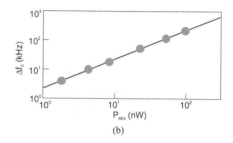

(a)       (b)

FIGURE 6.26  Frequency responsivity to absorbed light of graphene resonator: (a) mechanical resonance frequency *vs.* time for an 8-μm diameter trampoline with 500-nm wide tethers; the device is subject to 190 nW of incident radiation modulated at 40 Hz; (b) measured resonance shift *vs.* absorbed power, the absorbed power 4.4 nW causes a frequency shift of $\Delta f_0 = 8.5$ kHz (after Ref. [55]).

To drive motion of the graphene resonators, an AC voltage was applied between the graphene and the back-gate.

Upon absorbing light, the membrane's temperature increases and the resulting thermomechanical stress shifts the resonance frequency by an amount [55] of:

$$\Delta f_0 = \frac{\alpha Y f_0}{2\sigma_0 (1-\nu)} \Delta T,$$ (6.16)

where $\alpha$ is the thermal expansion coefficient, $\nu$ is the Poisson ratio, $\sigma_0$ is the initial in-plane stress, $Y$ is the two-dimensional elastic modulus, $f_0$ is the initial frequency, and $\Delta T$ is the temperature change. For typical graphene nanomechanical resonators, a $\Delta T \sim 100$ mK will shift the frequency by a full linewidth, a sizable amount that is readily measured.

Figure 6.26 shows the frequency responsivity to absorbed light by time recording of $f_0$ when the GNB is exposed to sinusoidally modulated light and the absorber power of 4.4 nW. A best-fit line to these data yields a 2.3 kHz/nW resonance shift per unit incident power.

The GNB device has achieved a sensitivity of 2 pW/Hz$^{1/2}$ and a bandwidth of up to 1 MHz, thus demonstrating a previously unattainable sensitivity at room temperature, and markedly outperforming the speed of state-of-the-art room-temperature bolometers.

## 6.4  GRAPHENE-BASED TERAHERTZ DETECTORS

Various detection mechanisms are involved in terahertz 2D-material detectors, including the bolometric effect, photothermoelectric (PTE) effect, plasma wave rectification in FET, and so on. Table 6.2 summarizes the performance of representative detectors [69]. The detectivity is estimated from Eq. (2.6), where *NEP* values are obtained from the references.

THz response can be attributed to the PTE effect as a result of asymmetric thermal distribution. Cai *et al.* reported a sensitive graphene PTE THz detector with asymmetric electrodes [58]. On account of dissimilar metal contacts, the Fermi energy and Seebeck coefficients are asymmetric across the device channel, which lead to a photovoltage in response to incident radiation. In addition, gate voltage may tune the Fermi energy of graphene, thus influencing the photoresponse. The detector response time is mainly determined by the thermal time constant of graphene, which is about 100 ps.

In addition, Tong *et al.* have demonstrated a half-edge-contacted graphene PTE detector with asymmetric electrodes (Fig. 6.27) [59]. Its sensitivity was enhanced by a double-patch antenna and an on-chip silicon lens, reaching a maximum responsivity of 4.9 V/W and a typical $D^*$ of $\sim 2.2 \times 10^6$ Jones.

Different modifications of detector design have been proposed to improve sensitivity. One of them is a resonant structure consisting of two graphene sheets, separated by a dielectric layer, to tune the absorbed wavelength, as presented in Table 6.1. The responsivity of that detector exhibits peaks when the frequency of inducing THz radiation reaches the resonant plasma frequencies being tuned by the bias [70]. In order to observe the resonant response of the detector, the frequency of electron collisions as a result of impurities and acoustic phonons must be sufficiently low. Double-graphene-layered (DGL) heterostructures, deploying inter-graphene layer intra-band transitions, show that such structures can be exploited for efficient and tunable THz/IR lasers and photodetectors [71].

The DGL heterostructure consists of a narrow potential barrier like BN and $WS_2$, sandwiched between two layers of graphene [Fig. 6.28(a)]. The voltage ($V_{\text{top-bottom}}$) applied between the source and the drain, combined with the gate voltage, $V_g$, modulates the Fermi levels and forms a p-i-n structure [Fig. 6.28(b)]. The system is in resonance when the energies of Dirac points of the two layers of graphene are exactly the same. The resonant tunneling between the two layers of graphene, with conservation of energy and momentum, results in the negative differential conductance. A photocurrent is generated due to band-offset ($\Delta$) between two Dirac points ($\hbar\omega$), and the TM-polarized photons with energy $\hbar\omega \sim \Delta$ can mediate the

TABLE 6.2 Terahertz Detectors Based on 2D Materials (after Ref. [69])

| Mechanism | Material | Frequency | Responsivity | NEP | $D^*$ | Response Time | T | Ref. |
|---|---|---|---|---|---|---|---|---|
| Bolometer | bP | ~0.3 THz | 7.8 V/W | 4 nW/Hz$^{1/2}$ | ~$1.2\times10^7$ Jones | <1 ms | 300 K | [56] |
| Bolometer | Graphene | 0.3–1.6 THz | – | 5.6 pW/Hz$^{1/2}$ | ~$3.3\times10^9$ Jones | <100 µs | 3 K | [57] |
| Bolometer | Graphene | 0.15 THz | $5\times10^{10}$ V/W | 0.2 fW/Hz$^{1/2}$ | ~$2.2\times10^{12}$ Jones | <2.5 ns | 300 K | [53] |
| PTE | Graphene | 2.52 THz | >10 V/W | 1.1 nW/Hz$^{1/2}$ | ~$1.9\times10^5$ Jones | <110 ps | 300 K | [58] |
| PTE | Graphene | 2 THz | >4.9 V/W | 1.7 nW/Hz$^{1/2}$ | ~$2.2\times10^6$ Jones | <50 ms | 300 K | [59] |
| PTE | bP | ~0.3 THz | >0.15 V/W | 40 nW/Hz$^{1/2}$ | ~$1.2\times10^6$ Jones | <2 ms | 300 K | [60] |
| PTE | bP | ~0.3 THz | >1.1 V/W | 45 nW/Hz$^{1/2}$ | ~$1.0\times10^6$ Jones | <2 ms | 300 K | [56] |
| FET | Graphene | ~0.3 THz | >0.15 V/W | 30 nW/Hz$^{1/2}$ | ~$1.5\times10^6$ Jones | <1 ms | 300 K | [61] |
| FET | Graphene | ~0.6 THz | >14 V/W | 515 pW/Hz$^{1/2}$ | ~$4.9\times10^7$ Jones | <30 µs | 300 K | [62] |
| FET | Graphene | ~0.3 THz | >1.2 V/W | 2 nW/Hz$^{1/2}$ | ~$2.3\times10^7$ Jones | <2.5 ms | 300 K | [63] |
| FET | Graphene | ~0.3 THz | >30 V/W | 163 pW/Hz$^{1/2}$ | ~$3.0\times10^8$ Jones | <5 µs | 300 K | [64] |
| FET | bP | ~0.3 THz | >5 V/W | 10 nW/Hz$^{1/2}$ | ~$4.8\times10^6$ Jones | <10 ms | 300 K | [56] |
| FET | Graphene | 230–375 GHz | >0.25 V/W | 10 nW/Hz$^{1/2}$ | ~$5.8\times10^5$ Jones | <1.2 ms | 300 K | [65] |
| FET | Graphene | 0.13 GHz | >20 V/W | 0.6 nW/Hz$^{1/2}$ | ~$1.9\times10^8$ Jones | – | 300 K | [66] |
| FET | bP | 0.15 THz | >300 V/W | 1 nW/Hz$^{1/2}$ | ~$1.7\times10^6$ Jones | <4 µs | 300 K | [67] |
| FET | Graphene | 0.15 THz | >400 V/W | 0.5 nW/Hz$^{1/2}$ | ~$3.5\times10^8$ Jones | <20 µs | 300 K | [68] |

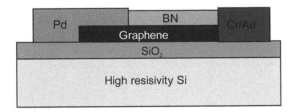

FIGURE 6.27   Schematic side-view of the THz PTE detector design.

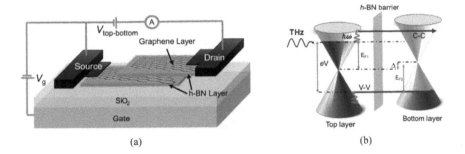

FIGURE 6.28   Double-graphene-layered (DGL) heterostructure: (a) schematic of the fabricated DGL device, (b) band diagrams of the DGL detector structures with photo-absorption-assisted inter-GL transitions. Wavy arrows indicate the inter-GL radiative V–V and C–C transitions. The inter-GL transitions work for the TM-mode THz photon radiations in the structure (after Ref. [71]).

electron photon-assisted resonant tunneling in two layers of graphene. Yadav *et al.* reported the first experimental observation of the double-graphene-layered heterostructure THz detector [71].

FET THz detectors conventionally operate at room temperature in the resistive self-missing regime, since the length of the FET channel exceeds the propagation length of plasma waves at room temperature. The plasma waves, excited by THz radiation, could be overdamped, and thus the detectors did not operate in the resonant regime, as was demonstrated by Vicarelli *et al.* [61]. The first THz FET detector, based on monolayer and bilayer graphene, was exfoliated on $Si/SiO_2$. In order to satisfy the boundary condition, the ends of the logarithmic antenna were the gate and source, whereas the drain was a metal line leading to a small pad. Figure 6.29(a) shows plasma wave FET with a top-gate antenna-coupled configuration. These detectors demonstrated a noise equivalent power (*NEP*) ~ $10^{-9}$ W/$Hz^{1/2}$ in the range 0.29–0.38 THz. It was reported that, during measurements of a target at room temperature, the bilayer graphene-based FET

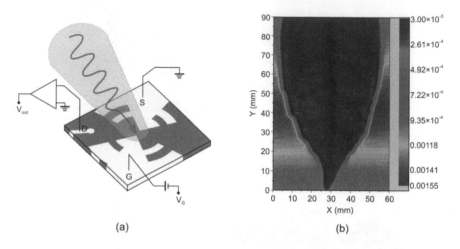

(a)                                                    (b)

FIGURE 6.29    Plasma wave FET THz detector: (a) THz detection configuration in a FET embedding the optical image of the central area of a bilayer graphene-based FET and (b) 0.3 GHz transmission mode image of a leaf (adapted after Ref. [61]).

at gate voltage, $V_g = 3$ V, was mounted on a x-y translation stage, having a spatial resolution of 0.5-μm [61]. The THz image consists of $200 \times 550$ scanned points collected by the object raster scanning in the beam focus, with integration time of 20 ms/point [Fig. 6.29(b)]. Further studies carried out by Spirito *et al.* [63] revealed the influence of the gate configuration in a bilayer graphene FET detector. The device exhibited better performance by employing wide-gate geometries or buried gate configurations [Fig. 6.30(b)], where a responsivity of 1.2 V/W was achieved.

More recently, improvements in THz detection have been described by Liu *et al.* [68,72]. The detector was built from split-finger gated

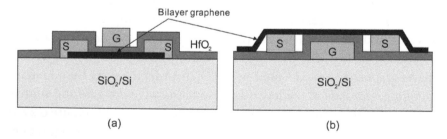

(a)                                                    (b)

FIGURE 6.30    Schematic bilayer graphene FET with (a) top and (b) buried gates (after Ref. [63])

graphene-based FET, coupled with a logarithmic antenna. Its *NEP* value, determined by the Johnson-Nyquist noise, was less than 0.1 nW/Hz$^{1/2}$ at both 0.04 and 0.14 THz. The achieved performances are competitive with those of commercially available detectors, in terms of both sensitivity and *NEP*.

In comparison with plasmonic THz detectors, fabricated by exfoliated graphene [63] or chemical vapor deposition (CVD) deposited on Si/SiO$_2$ substrates [62], the epitaxial graphene, grown on SiC, is much more promising [72]. The photoresponsivity and *NEP* of the bilayer graphene (deposited on SiC substrate) FET channel were estimated to be at the levels of 0.25 V/W and 80 nW/Hz$^{1/2}$, respectively.

## 6.5 GRAPHENE-BASED DETECTOR PERFORMANCE – THE PRESENT STATUS

Since its discovery, graphene has been extensively studied for its potential service as photodetector in a wide range of the electromagnetic spectrum, where the majority of research has been devoted to visible and NIR photon detectors [3,17,19,69]. The performance of the photodetector is mainly dependent on the inherent characteristics of its active layers, such as absorption coefficient, e-h pair lifetime, and charge mobility. The high dark current of conventional graphene materials, arising from the gapless nature of graphene, significantly reduces the photodetector's sensitivity and restricts further development of graphene-based photodetectors.

Published data on the performance of graphene-based infrared photodetectors are collated in Table 6.3 [19,21].

The spectral responsivity of graphene photodetectors, operating in the visible and NIR spectral ranges, are compared with commercially available silicon and InGaAs photodiodes in Fig. 6.31. The experimental results are taken from Ref. [73] and many other papers. The highest current responsivity, above 10$^7$ A/W, was achieved for hybrid Gr/quantum dot (QD) photodetectors with enhancement trapped charge lifetimes. As is shown, graphene's high mobility, together with the enhancement trapped charge lifetimes in the quantum dots, produced photodetector responsivity up to seven orders of magnitude higher than that of standard bulk photodiodes, where $g \approx 1$. The higher responsivity of the Si avalanche photodiode (APD), up to 100 A/W, is caused by the avalanche process. The high responsivity allows for fabrication of devices suitable for measuring low-level signals. However, due to the long lifetime of the traps, the

TABLE 6.3 Infrared Photoresponsivity of Graphene-based Materials

| Materials | Wavelength | Responsivity (A/W) | Bias (V) | Gain (%) | Time | Detectivity (Jones) |
|---|---|---|---|---|---|---|
| Graphene | 1 mm (0.3 THz) | 0.07–0.15 V/W | 0 | | | |
| Graphene | 30–220 $\mu$m | $(5-10)\times10^{-9}$ | 0.1 | | 10 ps, 50 ps | |
| Graphene | 1550 nm | $0.5\times10^{-3}$ | 1.5 | | 0.26 s | $3.3\times10^{13}$ |
| Graphene/Si waveguide | 2750 nm | 0.13 | | | | |
| PbSe/TiO$_2$/graphene | 350–1700 nm | 0.506–0.13 | –1 | | 50 ns, 83 ns | $3\times10^{13}$ |
| Graphene/PbS | 350–1700 nm | $5\times10^7$ | 5 | $10^8$ | 10–20 ms | $7\times10^{13}$ |
| Graphene/Ti$_2$O$_3$ | 3–13 $\mu$m | ~ 100 | | | | |

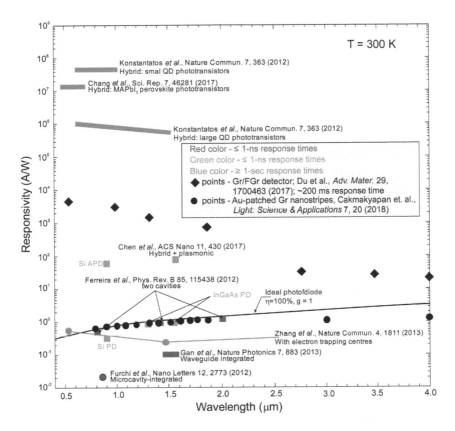

FIGURE 6.31 Spectral responsivity of graphene photodetectors compared with commercial ones. Solid line shows 100% quantum efficiency. Red and green colors denote ≤ 1-ns response times, while the blue color denotes ≥ 1-s response times. The graphene photodetectors are labeled with their reference as well as a brief description of the photodetector style. The commercial photodiodes are shown in green.

demonstrated frequency response of 2D material photodetectors is very slow (<10 Hz), which considerably limits real detector functions.

It is interesting to highlight the unique electrical and optical characteristics of gold-patched graphene nanostripe photodetectors demonstrated by Cakmakyapan *et al.* [48]. The photodetector has an ultrabroad spectral response from the visible to the IR region, with high responsivity levels ranging from 0.6 A/W at a wavelength of 800 nm to 11.65 A/W at 20 μm, with an operation speed exceeding 50 GHz. As shown in Fig 6.31, its current responsivity (black circles) coincides well with the curve (black line) theoretically predicted for an ideal photodiode in the near-infrared (NIR) spectral range.

As with graphene-based photodetectors, ultrahigh responsivity and ultrashort time response cannot be obtained at the same time in practice with 2D-related material photodetectors. The 2D-layered materials show potential in photodetection covering the UV, visible, and IR ranges (Fig. 6.32). Generally, however, most of them cover only the visible and NIR spectral range, and only graphene-based, bPAs, bismuthene (like $Bi_2Te_3$ and $Bi_2Se_3$) and noble transition metal dichalcogenides play a major role in the infrared region. In addition, as Fig. 6.32 shows, the performance of graphene-based infrared photodetectors is inferior in comparison with alternative 2D-material photodetectors.

Published data for longer-wavelength infrared graphene-based photodetectors, with a cutoff wavelength above 3 μm, are limited. Figure 3.15 compares their detectivities with commercially available HgCdTe photodiodes [76] and type-II InAs/GaSb superlattice inter-band quantum cascade infrared photodetectors (IB QCIPs) [77], operating at room temperature. The upper detectivity of Gr/FGr photodetector in SWIR range is close to that of HgCdTe photodiodes. This figure also shows experimental data for black phosphorus arsenic (bAsP) photodetectors and noble transition metal dichalcogenides entering the second atmospheric transmission window.

Figure 3.16 compares the experimental detectivity values published in the literature for different types of single-element 2D-material photodetectors, operating at room temperature, with theoretically predicted

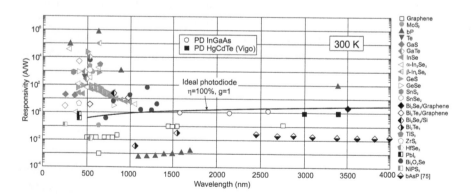

FIGURE 6.32 Summary of spectral responsivities to the layered 2D material photodetectors at room temperature (after Refs. [74,75]). Black line shows spectral responsivity for ideal photodiode with 100% quantum efficiency and $g=1$. For comparative goals, the responsivities of commercially available photodetectors (InGaAs and HgCdTe photodiodes) are marked.

curves for P-i-N HOT HgCdTe photodiodes. As is shown, the detectivity values for selected 2D-material photodetectors are close to the data presented for commercial detectors ((PV Si and Ge, PV InGaAs, PC PbS and PbSe, PV HgCdTe), and, in the case of black phosphorus and noble TMD detectors, are even higher. The enhanced sensitivity of 2D-material photodetectors is achieved by bandgap engineering and photogating effects. However, the layered-material photodetectors are characterized by limited linear dynamic range of operation and slow response times.

Particular attention in development of THz imaging systems is paid to the fabrication of sensors with a high potential for real-time imaging, while maintaining a high dynamic range and room-temperature operation. CMOS process technology is especially attractive, due to their low price for industrial, surveillance, scientific, and medical applications. With that in mind, much of the recent development has been directed towards three types of focal plane arrays (FPAs):

- Schottky barrier diodes (SBDs), compatible with the CMOS process,
- FETs relying on the plasmonic rectification effect, and
- Adaptation of IR bolometers to the THz frequency range.

SPDs respond to the THz electric field and usually generate an output current or voltage through a quadratic term in their current-voltage characteristics. In general, the *NEP* of SBD and FET detectors is better than that of Golay cells and pyroelectric detectors, around 300 GHz. Both the pyroelectric and the bolometer FPAs with detector response times in the millisecond range are not suitable for heterodyne operation. FET detectors are clearly capable of heterodyne detection with improved sensitivity. Diffraction aspects predict FPAs for higher frequencies (0.5 THz and above) and in conjunction with large *f/#* optics.

Owing to its high carrier mobility, graphene is a very promising material for the development of room-temperature detectors operating across the FIR, with high room-temperature performance for high spectral bandwidth covering the full THz range (0.1–10 THz).

At the present stage of technology, the most effective graphene THz detectors employ the plasma rectification effect in FETs, where plasma waves in the channel are excited by incoming THz waves, modulating the potential difference between gate and source/drain, and being rectified *via* non-linear coupling and transfer characteristics in FETs. FETs indeed provide some advantages at THz frequencies, namely the inherent scalability and the combination of a fast response with high frequency operation. For comparison, the performances of Schottky diodes are strongly affected by parasitic capacitances and usually show a dramatic cutoff above 1 THz. Figures 6.33 and 6.34 compare the *NEP* values of graphene-based FET room-temperature detectors with existing THz photon detectors (Fig. 6.33) and thermal detectors (Fig. 6.34) dominating the market. The experimental data are gathered from literature world-wide.

Most experimental data gathered in the literature are given for single-graphene detectors operating above a wavelength of 100 μm (frequency range below 3 THz). Generally, the performance of graphene FET detectors is lower than those of CMOS-based and plasma detectors, fabricated using silicon-, SiGe-, and InGaAs-based materials. However, in comparison with $VO_x$ and amorphous silicon microbolometers, the performance of graphene detectors is close to the trend line estimated for microbolometers in the THz spectral region (Fig. 6.34). The best quality $VO_x$ bolometer arrays are characterized by *NEP* values of about 1 pW/Hz$^{1/2}$ in the LWIR range ($\approx$ 10 μm). It should be noted here, however, that microbolometer

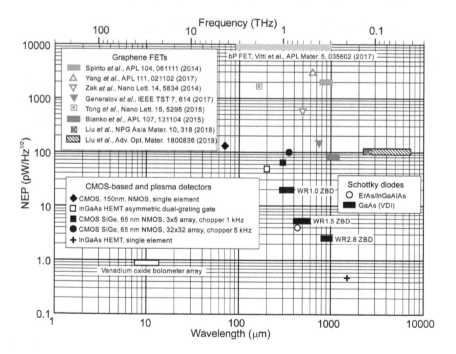

FIGURE 6.33 *NEP* spectral dependence for graphene FET detectors and different photon THz (CMOS-based, Schottky diodes).

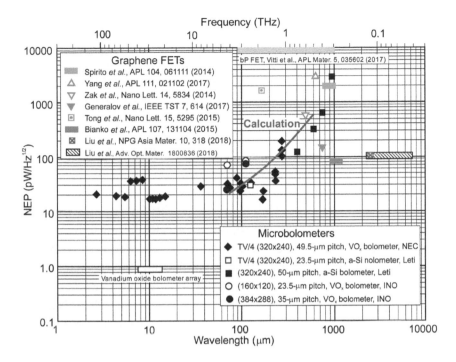

FIGURE 6.34 *NEP* spectral dependence for graphene FET detectors and microbolometer THz FPAs.

data are addressed to monolithic arrays. In this case, an important issue for FPA is pixel uniformity. It appears that the fabrication of monolithically integrated THz detector arrays encounters many technological problems, with the device-to-device performance variations and even the percentage of non-functional detectors per chip tending to be unacceptably high. After the successful demonstration of active THz imaging [78], adaptation of IR microbolometers to the THz frequency range meant that, in the period 2010–2011, three different companies/organizations announced cameras optimized for the terahertz frequency range: NEC (Japan) [79], INO (Canada) [80] and LETI (France) [81]. Figure 6.34 summarizes the *NEP* values for bolometer FPAs fabricated by each of the three vendors. The FPAs optimized for 2–5 THz exhibit impressive *NEP* values below 100 $pW/Hz^{1/2}$. It can be seen that the wavelength dependence of *NEP* is quite flat below 200 μm. Further improvement of performance is possible by increasing the number of pixels, or modifying the antenna design, while preserving pixel pitch, readout integrated circuit (ROIC), and technological stack.

As is shown in Section 4.3, for detector arrays [so-called focal plane arrays (FPAs)], the relevant figure of merit for determining the ultimate performance is not the detectivity, $D^*$, but the noise equivalent difference temperature (*NEDT*) and the modulation transfer function (*MTF*) [1]. *NEDT* and *MTF* reflect the primary performance metrics for thermal imaging systems: thermal sensitivity and spatial resolution, respectively. Thermal sensitivity is concerned with the minimum temperature difference that can be discerned above the noise level. The *MTF* concerns the spatial resolution and answers one question – how small an object can be imaged by the system?

In 2013, the European Union launched the Future and Emerging Technology Graphene Flagship program to accelerate research into technologies based on graphene and related materials. Recently, within the framework of this program, the hybrid graphene-colloidal quantum dot (QD) photodetector arrays, fully vertically integrated into a CMOS readout chip, have been demonstrated (Fig. 6.35). These 388×288 pixel cameras are operated in the UV-visible-SWIR range from 300 to 2000 nm. Pixels of the image sensor (Table 6.1, first line) are characterized by a high gain of $10^7$ and a responsivity above $10^7$ A/W. The size of the pixels is large (20-μm, which limits spatial resolution) for operation in the visible and SWIR range, in comparison with present commercial CMOS images operating in the visible region (Fig. 4.8). Furthermore, operability of the arrays, estimated as 95%, is poor [82]. The fixed pattern noise

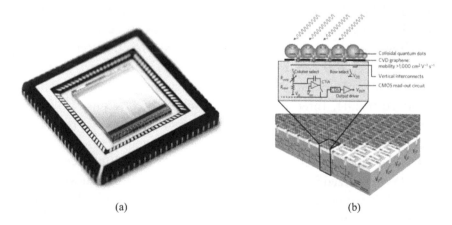

(a)                                         (b)

FIGURE 6.35   (a) Monolithically integrated graphene-quantum dot photodetector array and (b) side-view of detector and the underlying readout circuit (after Ref. [82]).

(spatial noise) has strong influence on the noise equivalent irradiance of the arrays, especially if the array's pixels are hybrid (graphene-QDs). It is well known that nonuniformity of QDs limits performance of QD photodetector arrays [83].

From the above discussion about the present status of single graphene-based detectors, we can conclude that:

- Pristine graphene has a broadband absorption and a fast response, dominated by the dynamics of hot carriers; in consequence, due to its low absorption coefficient and fast relaxation time, its responsivity is lower in comparison with detectors already on the global market, especially HgCdTe, InGaAs, and microbolometers.

- Improving the responsivity by combining graphene with other materials (hybrid photodetectors), because of the photogating effect, results in a limited linear dynamic range due to the charge relaxation time, leading to a drop in sensitivity with incident optical power.

- Responsivity of hybrid and chemically functionalized graphene detectors is comparable with and sometimes even better than that of commercially available detectors, although a considerable drop in operating speed (bandwidth) is observed; generally, their response time (millisecond range and longer) is three orders of magnitude longer in comparison with commercial detectors (microsecond range and shorter); the above drawbacks limit the practical applications of these hybrid and chemically functionalized graphene detectors.

- At the present stage of the technology, the most effective graphene-based photodetectors are the THz detectors which utilize plasma rectification phenomena in FETs; performance of these detectors approximates to that of commercial detectors.

- Due to poor graphene-based detector technology, development of arrays is in its infancy (especially in terms of very low operability and uniformity).

Considering the fundamental material properties of graphene and the present state of graphene technology, it is rather doubtful that a stronger position for graphene in the future infrared imaging systems can be predicted. So far, most device applications of graphene have been suggestions rather than demonstrated capabilities. The assumptions that graphene is always a perfect 2D crystal in isolation from its environment is, in most cases, not valid. There are significant challenges faced by materials scientists to improve graphene growth techniques.

Graphene-based detector challenges include the limited linear dynamic range of operation, the lack of efficient generation and extraction of photoexcited charges, the smearing of photoactive junctions due to hot-carrier effects, large-scale fabrication, and ultimately the environmental stability of the constituent materials [84].

The ratio of the absorption coefficient to the thermal generation rates, $\alpha/G$, is the main figure of merit of any IR material, where the thermal generation rate is inversely proportional to the recombination lifetime [85]. Graphene is an attractive material for optical detection due to its broad absorption spectrum and ultrashort response time. Its ultrahigh mobility is suitable for high-speed communications. However, it remains a great challenge to achieve high responsivity in graphene detectors because of graphene's weak optical absorption (only 2.3% in the monolayer graphene sheet) and short carrier lifetime (<1 ps). In other words, the applications of graphene-based photodetectors are limited in comparison with traditional detectors. Various approaches have been proposed to enhance sensitivity by introduction of a bandgap, electron trap layers (quantum dot structures), or nanoribbons. However, these methods degrade the electronic performance, including the ever-important carrier mobility.

## REFERENCES

1. A. Rogalski, *Infrared and Terahertz Detectors*, 3rd edition, CRC Press, Boca Raton, 2019.
2. G. Konstantatos, M. Badioli, L. Gaudreau, J. Osmond, M. Bernechea, F.P. Garcia de Arquer, F. Gatti, and F.H.L. Koppens "Hybrid graphene-quantum dot phototransistors with ultrahigh gain", *Nature Nanotechnology* **7**, 363–368 (2012).
3. F.H.L. Koppens, T. Mueller, Ph. Avouris, A.C. Ferrari, M.S. Vitiello and M. Polini, "Photodetectors based on graphene, other two-dimensional materials and hybrid systems", *Nature Nanotechnology* **9**, 780–793, 2014.
4. H. Fang and W. Hu, "Photogating in low dimensional photodetectors", *Advanced Science* **4**, 1700323 (2017).
5. P. Wang, H. Xia, Q. Li, F. Wang, L. Zhang, T. Li, P. Martyniuk, A. Rogalski, and W. Hu, "Sensing infrared photons at room temperature: from bulk materials to atomic layers", *Small* **46**(13), 1904396 (2019).
6. X. Li, L. Tao, Z. Chen, H. Fang, X. Li, X. Wang, J.-B. Xu, and H. Zhu, "Graphene and related two-dimensional materials: Structure-property relationships for electronics and optoelectronics", *Applied Physics Reviews* **4**, 021306-1–31 (2017).

7. N. Mott and H. Jones, *The Theory of the Properties of Metals and Alloys*, The Clarendon Press, Oxford, 1936.

8. A. Nourbakhsh, L. Yu, Y. Lin, M. Hempel, R.-J. Shiue, D. Englund, and T. Palacios, "Heterogeneous integration of 2D materials and devices on a Si platform", in *Beyond-CMOS Technologies for Next Generation Computer Design*, pp. 43–84, ed. R.O. Topaloglu and H.-S.P. Wong, Springer, 2019.

9. T. Low and P. Avouris, "Graphene plasmonic for terhertz to mid-infrared applications", *ACS Nano* **8**(2), 1086–1001 (2014).

10. F. Xia, H. Yan, and P. Avouris, "The interaction of light and graphene: Basic, devices, and applications", *Proceedings of IEEE* **101**(7), 1717–1731 (2013).

11. X. Du, D.E. Prober, H. Vora, and Ch.B. Mckitterick, "Graphene-based bolometers", *Graphene and 2D Materials* **1**, 1–22 (2014).

12. E. Pop, V. Varshney, and A.K. Roy, "Thermal properties of graphene: Fundamentals and applications", *MRS Bulletin* **37**, 1273–1281 (2012).

13. M. Dyakonov and M.S. Shur, "Shallow water analogy for a ballistic field effect transistor: new mechanism of plasma wave generation by the dc current", *Physical Review Letters* **71**, 2465–2468 (1993).

14. M. Dyakonov and M. Shur, "Plasma wave electronics: Novel terahertz devices using two dimensional electron fluid", *IEEE Transactions on Electron Devices* **43**, 1640–1646 (1996).

15. M. Shur and V. Ryzhii, "Plasma wave electronics", *International Journal of High Speed Electronics and Systems* **13**, 575–600 (2003).

16. H.-X. Wang, Q. Wang, K.-G. Zhou, and H.-L. Zhang, "Graphene in light: Design, synthesis and applications of photo-active graphene and graphene-like materials", *Small* **9**(8), 1266–1283 (2013).

17. M. Buscema, J.O. Island, D.J. Groenendijk, S.I. Blanter, G.A. Steele, H.S.J. van der Zant, and A. Castellanos-Gomez, "Photocurrent generation with two-dimensional van der Waals semiconductor", *Chemical Society Reviews* **44**, 3691–3718 (2015).

18. G. Konstantatos, "Current status and technological prospect of photodetectors based on two-dimensional materials", *Nature Communications* **9**, 52661–52663 (2018).

19. M. Long, P. Wang, H. Fang, and W. Hu, "Progress, challenges, and opportunities for 2D material based photodetectors", *Advanced Functional Materials* **29**, 1803807 (2018).

20. Y. Zhang, T. Liu, B. Meng, X. Li, G. Liang, X. Hu, and Q.J. Wang, "Broadband high photoresponse from pure monolayer graphene photodetector", *Nature Communications* **4**, 1811 (2013).

21. D. Chronopoulos, A. Bakandritsos, M. Pykal, R. Zboril, and M. Otyepka, "Chemistry, properties, and applications of fluorographene", *Applied Materials Today* **9**, 60–70 (2017).

22. Z. Sun, Z. Liu, J. Li, G.-an Tai, S.-P. Lau, and F. Yan, "Infrared photodetectors based on CVD-grown graphene and PbS quantum dots with ultrahigh responsivity", *Advanced Materials* **24**, 5878–5883 (2012).

23. W. Guo, S. Xu, Z. Wu, N. Wang, M.M.T. Loy, and S. Du, "Oxygen-assisted charge transfer between ZnO quantum dots and graphene", *Small* **9**, 3031–3036 (2013).

24. I. Nikitskiy, S. Goossens, D. Kufer, T. Lasanta, G. Navickaite, F.H. Koppens, and G. Konstantatos, "Integrating an electrically active colloidal quantum dot photodiode with a graphene phototransistor", *Nature Communications* **7**, 11954 (2016).

25. X. Yu, Y. Li, X. Hu, D. Zhang, Y. Tao, Z. Liu, Y. He, M.A. Haque, Z. Liu, T. Wu, and Q.J. Wang, "Narrow bandgap oxide nanoparticles coupled with graphene for high performance mid-infrared photodetection", *Nature Communications* **9**, 4299 (2018).

26. S. Du, W. Lu, A. Ali, P. Zhao, K. Shehzad, H. Guo, L. Ma, X. Liu, X. Pi, P. Wang, H. Fang, Z. Xu, C. Gao, Y. Dan, P. Tan, H. Wang, C.-T. Lin, J. Yang, S. Dong, Z. Cheng, E. Li, W. Yin, J. Luo, B. Yu, T. Hasan, Y. Xu, W. Hu, and X. Duan, "A broadband fluorographene photodetector", *Advanced Materials* **29**, 1700463 (2017).

27. V. Ryzhii, T. Otsuji, M. Ryzhii, and M.S. Shur, "Double graphene-layer plasma resonances terahertz detector", *Journal of Physics D: Applied Physics* **45**, 302001 (6pp) (2012).

28. C.H. Liu, Y.C. Chang, T.B. Norris, and Z.H. Zhong, "Graphene photodetectors with ultra-broadband and high responsivity at room temperature", *Nature Nanotechnology* **9**, 273–278 (2014).

29. X. Gan, R.-J. Shiue, G. Yuanda, I. Meric, T.F. Heinz, K. Shepard, J. Hone, S. Assefa and D. Englund, "Chip-integrated ultrafast graphene photodetector with high responsivity", *Nature Photonics* **7**, 883–887 (2013).

30. A. Pospischil, M. Humer, M.M. Furchi, D. Bachmann, R. Guider, T. Fromherz, and T. Mueller, "CMOS-compatible graphene photodetector covering all optical communication bands", *Nature Photonics* **7**, 892–896 (2013).

31. X.M. Wang, Z.Z. Cheng, K. Xu, H.K. Tsang, and J.B. Xu, "High-responsivity graphene/silicon-heterostructure waveguide photodetectors", *Nature Photonics* **7**, 888–891 (2013).

32. D. Schall, D. Neumaier, M. Mohsin, B. Chmielak, J. Bolten, C. Porschatis, A. Prinzen, C. Matheisen, W. Kuebart, B. Junginger, W. Templ, A.L. Giesecke, and H. Kurz, "50 GBit/s photodetectors based on wafer-scale graphene for integrated silicon photonic communication systems", *ACS Photonics* **1**, 781–784 (2014).

33. T.J. Echtermeyer, L. Britnell, P.K. Jasnos, A. Lombardo, R.V. Gorbachev, A.N. Grigorenko, A.K. Geim, A.C. Ferrari, and K.S. Novoselov, "Strong plasmonic enhancement of photovoltage in graphene", *Nature Communications* **2**, 458 (2011).

34. Y. Liu, R. Cheng, L. Liao, H. Zhou, J. Bai, G. Liu, L. Liu, Y. Huang, and X. Duan, "Plasmon resonance enhanced multicolour photodetection by graphene", *Nature Communications* **2**, 579 (2011).

35. D.A. Bandurin, D. Svintsov, I. Gayduchenko, S.G. Xu, A. Principi, M. Moskotin, I. Tretyakov, D. Yagodkin, S. Zhukov, T. Taniguchi, K. Watanabe, I.V. Grigorieva, M. Polini, G.N. Goltsman, A.K. Geim, and G. Fedorov, "Resonant terahertz detection using graphene plasmons", *Nature Communications* **9**, 5392-1–8 (2018).

36. M. Engel, M. Steiner, A. Lombardo, A.C. Ferrari, H.V. Löhneysen, P. Avouris, and R. Krupke, "Light-matter interaction in a microcavity-controlled graphene transistor", *Nature Communications* **3**, 906 (2012).
37. X. Gan, K.F. Mak, Y. Gao, Y. You, F. Hatami, J. Hone, T.F. Heinz, and D. Englund, "Strong enhancement of light-matter interaction in graphene coupled to a photonic crystal nanocavity", *Nano Letters* **12**, 5626–5631 (2012).
38. M. Furchi, A. Urich, A. Pospischil, G. Lilley, K. Unterrainer, H. Detz, P. Klang, A.M. Andrews, W Schrenk, G. Strasser, and T. Mueller, "Microcavity-integrated graphene photodetector", *Nano Letters* **12**, 2773–2777 (2012).
39. M. Freitag, T. Low, W. Zhu, H. Yan, F. Xia, and P. Avouris, "Photocurrent in graphene harnessed by tunable intrinsic plasmons", *Nature Communications* **4**, 1951 (2013).
40. S. Ogawa, M. Shimatani, S. Fukushima, S. Okuda, and K. Matsumoto, "Graphene on metal-insulator-metal-based plasmonic metamaterials at infrared wavelengths", *Optics Express* **26**(5), 5665–5674 (2018).
41. R.-J. Shiue, X. Gan, Y. Gao, L. Li, X. Yao, A. Szep, D. Walker Jr., J. Hone, and D. Eglund, "Enhanced photodetection in graphene-integrated photonic crystal cavity", *Applied Physics Letters* **103**, 241109 (2013).
42. Z. Fang, Z. Liu, Y. Wang, P.M. Ajayan, P. Nordlander, and N.J. Halas, "Graphene–antenna sandwich photodetector", *Nano Letters* **12**, 3808–3813 (2012).
43. Y. Yao, R. Shankar, P. Rauter, Y. Song, J. Kong, M. Loncar, and F. Capasso, "High-responsivity mid-infrared graphene detectors with antenna enhanced photocarrier generation and collection", *Nano Letters* **14**, 3749–3754 (2014).
44. F. Liu, H. Shimotani, H. Shang, T. Kanagasekaran, V. Zolyomi, N. Drummond, V.I. Fal'ko, and K. Tanigaki, "High-sensitivity photodetectors based on multilayer GaTe flakes", *ACS Nano* **8**(1), 752–760 (2014).
45. T.J. Echtermeyer, S. Milana, U. Sassi, A. Eiden, M. Wu, E. Lidorikis, and A.C. Ferrari, "Surface plasmon polariton graphene photodetectors", *Nano Letters* **16**, 8–20 (2015).
46. T. Liu, L. Tong, X. Huang, and L. Ye, "Room-temperature infrared photodetectors with hybrid structure based on two-dimensional materials", *Chinese Physics B* **28**(1), 017302 (2019).
47. Z. Chen, X. Li, J. Wang, L. Tao, M. Long, S.-J. Liang, L.K. Ang, C. Shu, H.K. Tsang, and J.-B. Xu, "Synergistic effects of plasmonics and electron trapping in graphene short-wave infrared photodetectors with ultrahigh responsivity", *ACS Nano* **11**, 430–437 (2017).
48. S. Cakmakyapan, P.K. Lu, A. Navabi, and M. Jarrahi, "Gold-patched graphene nano-stripes for high-responsivity and ultrafast photodetection from the visible to infrared regime", *Light: Science and Applications* **7**, 20-1–9 (2018).
49. C. Peng, S. Nanot, R.-J. Shiue, G. Grosso, Y. Yang, M. Hempel, P. Jarillo-Herrero, J. Kong, F.H.L. Koppens, D.K. Efetov, and D. Englund, "Compact mid-infrared graphene thermopile enabled by a nanopatterning technique of electrolyte gates", *New Journal of Physics* **20**, 083050 (2018).

50. A.L. Hsu, P.K. Herring, N.M. Gabor, S. Ha, Y.C. Shin, Y. Song, M. Chin, M. Dubey, A.P. Chandrakasan, J. Kong, P. Jarillo-Herrero, and T. Palacios, "Graphene-based thermopile for thermal imaging applications", *Nano Letters* **15**, 7211–7216 (2015).

51. J. Yan, M.-H. Kim, J.A. Elle, A.B. Sushkov, G.S. Jenkins, H.M. Milchberg, M.S. Fuhrer, and H.D. Drew, "Dual-gated bilayer graphene hot electron bolometer", *Nature Nanotechnology* **7**(7), 472–478, 2012.

52. U. Sassi, R. Parret, S. Nanot, M. Bruna, S. Borini, D. De Fazio, Z. Zhao, E. Lidorikis, F.H.L. Koppens, A.C. Ferrar, and A. Colli, "Graphene-based mid-infrared room-temperature pyroelectric bolometers with ultrahigh temperature coefficient of resistance", *Nature Communications* **8**, 1431 (2017).

53. A. El Fatimy, R.L. Myers-Ward, A.K. Boyd, K.M. Daniels, D.K. Gaskill, and P. Barbara, "Epitaxial graphene quantum dots for high-performance THz bolometers", *Nature Nanotechnology* **11**, 335–338 (2016).

54. A. El Fatimy, A. Nath, B.D. Kong, A.K. Boyd, R.L. Myers-Ward, K.M. Daniels, M.M. Jadidi, T.E. Murphy, D.K. Gaskill, and P. Barbara, "Ultra-broadband photodetectors based on epitaxial graphene quantum dots", *Nanophotonics* **7**(4), 735–740 (2018).

55. A. Blaikie, D. Miller, and B.J. Alemán, "A fast and sensitive room-temperature graphene nanomechanical bolometer", *Nature Communications* **10**, 4726 (2019).

56. L. Viti, J. Hu, D. Coquillat, A. Politano, W. Knap, and M.S. Vitiello, "Efficient terahertz detection in black-phosphorus nano-transistors with selective and controllable plasma-wave, bolometric and thermoelectric response", *Science Reports* **6**, 20474 (2016).

57. W. Miao, H. Gao, Z. Wang, W. Zhang, Y. Ren, K.M. Zhou, S.C. Shi, C. Yu, Z.Z. He, Q.B. Liu, and Z.H. Feng, "A graphene-based terahertz hot electron bolometer with Johnson noise readout", *Journal of Low Temperature Physics* **193**(3–4), 387–392 (2018).

58. X. Cai, A.B. Sushkov, R.J. Suess, M.M. Jadidi, G.S. Jenkins, L.O. Nyakiti, R.L. Myers-Ward, S. Li, J. Yan, D.K. Gaskill, T.E. Murphy, H.D. Drew, and M.S. Fuhrer, "Sensitive room-temperature terahertz detection via the photothermoelectric effect in graphene", *Nature Nanotechnology* **9**(10), 814–819 (2014).

59. J. Tong, M. Muthee, S.Y. Chen, S.K. Yngvesson, and J. Yan, "Antenna enhanced graphene THz emitter and detector", *Nano Letters* **15**(8), 5295–5301 (2015).

60. L. Viti, J. Hu, D. Coquillat, W. Knap, A. Tredicucci, A. Politano, and M.S. Vitiello, "Black phosphorus terahertz photodetectors", *Advanced Materials* **27**, 5567–5572 (2015).

61. L. Vicarelli, M.S. Vitiello, D. Coquillat, A. Lombardo, A.C. Ferrari, W. Knap, M. Polini, V. Pellegrini, and A. Tredicucci, "Graphene field-effect transistors as room temperature terahertz detectors", *Nature Materials* **11**(10), 865–871 (2012).

62. A. Zak, M.A. Andersson, M. Bauer, J. Matukas, A. Lisauskas, H.G. Roskos, and J. Stake, "Antenna-integrated 0.6 THz FET direct detectors based on CVD graphene", *Nano Letters* **14**(10), 5834–5838 (2014).

63. D. Spirito, D. Coquillat, S.L. De Bonis, A. Lombardo, M. Bruna, A.C. Ferrari, V. Pellegrini, A. Tredicucci, W. Knap, and M.S. Vitiello, "High performance bilayer graphene terahertz detectors", *Applied Physics Letters* **104**(6), 061111 (2014).

64. H. Qin, J. Sun, S. Liang, X. Li, X. Yang, Z. He, C. Yu, and Z. Feng, "Room-temperature, low-impedance and high-sensitivity terahertz direct detector based on bilayer graphene field-effect transistor", *Carbon* **116**, 760–765 (2017).

65. F. Bianco, D. Perenzoni, D. Convertino, S.L. De Bonis, D. Spirito, M. Perenzoni, C. Coletti, M.S. Vitiello, and A. Tredicucci, "Terahertz detection by epitaxial-graphene field-effect-transistors on silicon carbide", *Applied Physics Letters* **10**(13) 131104 (2015).

66. D.A. Bandurin, I. Gayduchenko, Y. Cao, M. Moskotin, A. Principi, I.V. Grigorieva, G. Goltsman, G. Fedorov, and D. Svintsov, "Dual origin of room temperature sub-terahertz photoresponse in graphene field effect transistors", *Applied Physics Letters* **112**(14) 141101 (2018).

67. L. Wang, C. Liu, X. Chen, J. Zhou, W. Hu, X. Wang, J. Li, W. Tang, A. Yu, S.-W. Wang, and W. Lu, "Toward sensitive room-temperature broadband detection from infrared to terahertz with antenna-integrated black phosphorus photoconductor", *Advanced Functional Materials* **27**(7), 1604414 (2017).

68. C. Liu, L. Wang, X. Chen, J. Zhou, W. Hu, X. Wang, J. Li, Z. Huang, W. Zhou, W. Tang, G. Xu, S.-W. Wang, and W. Lu, "Room-temperature photoconduction assisted by hot-carriers in graphene for sub-terahertz detection", *Carbon* **130**, 233–240 (2018).

69. Y. Wang, W. Wu, and Z. Zhao, "Recent progress and remaining challenges of 2D material-based terahertz detectors", *Infrared Physics and Technology* **102**, 103024 (2019).

70. V. Ryzhii, T. Otsuji, M. Ryzhii, and M.S. Shur, "Double graphene-layer plasma resonances terahertz detector", *Journal of Physics D: Applied Physics* **45**, 302001 (2012).

71. D. Yadav, S.B. Tombet, T. Watanabe, S. Arnold, V. Ryzhii, and T. Otsuji, "Terahertz wave generation and detection in double-graphene layered van der Waals heterostructures", *2D Materials* **3**(4), 045009 (2016).

72. C. Liu, L. Wang, X. Chen, A. Politano, D. Wei, G. Chen, W. Tang, W. Lu, and A. Tredicucci, "Room-temperature high-gain long-wavelength photodetector via optical-electrical controlling of hot carriers in graphene", *Advanced Optical Materials* **6**(24), 1800836 (2018).

73. M. Currie, "Applications of graphene to photonics", NRL/MR/5650-14-9550, 2014.

74. F. Wang, Z. Wang, L. Yin, R. Cheng, J. Wang, Y. Wen, T.A. Shifa, F. Wang, Y. Zhang, X. Zhan, and J. He, "2D library beyond graphene and transition metal dichalcogenides: a focus on photodetection", *Chemical Society Reviews* **47**(16), 6296–6341 (2018).

75. M. Long, A. Gao, P. Wang, H. Xia, C. Ott, C. Pan, Y. Fu, E. Liu, X. Chen, W. Lu, T. Nilges, J. Xu, X. Wang, W. Hu, and F. Miao, "Room temperature high-detectivity mid-infrared photodetectors based on black arsenic phosphorus", *Science Advances* **3**, e1700589 (2017).

76. https://vigo.com.pl/wp-content/uploads/2017/06/VIGO-Catalogue.pdf
77. A. Rogalski, P. Martyniuk, and M. Kopytko, "Type-II superlattice photodetectors versus HgCdTe photodiodes", *Progress in Quantum Electronics* **68**, 100228 (2019).
78. A.W.M. Lee, B.S.Williams, S. Kumar, Q. Hu, and J.L. Reno, "Real-time imaging using a 4.3-THz quantum cascade laser and a 320×240 microbolometer focal-plane array", *IEEE Photonics Technology Letters* **18**, 1415–1417 (2006).
79. N. Oda, "Uncooled bolometer-type terahertz focal-plane array and camera for real-time imaging", *Comptes Rendus Physique* **11**, 496–509 (2010).
80. M. Bolduc, M. Terroux, B. Tremblay, L. Marchese, E. Savard, M. Doucet, H. Oulachgar, C. Alain, H. Jerominek, and A. Bergeron, "Noise-equivalent power characterization of an uncooled microbolometer-based THz imaging camera", *Proceedings of SPIE* **8023**, 80230C–1–10 (2011).
81. D.-T. Nguyen, F. Simoens, J.-L. Ouvrier-Buffet, J. Meilhan, and J.-L. Coutaz, "Broadband THz uncooled antenna-coupled microbolometer array—electromagnetic design, simulations and measurements. *IEEE Transactions on Terahertz Science and Technology* **2**, 299–305 (2012).
82. S. Goossens, G. Navickaite, C. Monasterio, S. Gupta, J.J. Piqueras, R. Pérez, G. Burwell, I. Nikitskiy, T. Lasanta, T. Galán, E. Puma, A. Centeno, A. Pesquera, A. Zurutuza, G. Konstantatos, and F. Koppens, "Broadband image sensor array based on graphene–CMOS integration", *Nature Photonics* **11**, 366–371 (2017).
83. J. Phillips, "Evaluation of the fundamental properties of quantum dot infrared detectors", *Journal of Applied Physics* **91**, 4590–4594 (2002).
84. A. De Sanctis, J.D. Mehew, M.F. Craciun, and S. Russo, "Graphene-based light sensing: Fabrication, characterisation, physical properties and performance", *Materials* **11**, 1762 (2018).
85. J. Piotrowski and A. Rogalski, Comment on "Temperature limits on infrared detectivities of InAs/In$_x$Ga$_{1-x}$Sb superlattices and bulk Hg$_{1-x}$Cd$_x$Te" [*Journal of Applied Physics.* **74**, 4774 (1993)], *Journal of Applied Physics.* **80**(4), 2542–2544 (1996).

# CHAPTER 7

# Related 2D-Material Detectors

G RAPHENE IS ONE OF a large number of possible 2D crystals. The discovery of new 2D materials, with direct energy gaps in the IR to the
visible spectral regions, has opened up a new perspective on photodetector
fabrication. There are hundreds of layered materials that retain their stability down to constituent monolayers, the properties of which are complementary to those of graphene, and which have been covered in review
papers [1–9]. More information on this topic is included in Chapter 5.

## 7.1 GENERAL OVERVIEW

Detection mechanisms in 2D-material photodetectors, as for graphene-
based detectors (Chapter 6), have been reported, including the photoconductive, photovoltaic, photothermoelectric, and field-effect transistors
(FET) effects. For example, Fig. 7.1 shows two types of detectors: the $MoSe_2$
photoconductor and the black phosphorus (bP)-based FET. Whereas 2D
transition metal dichalcogenides (TMDs) are limited to UV-NIR, due to
their bandgaps, bP can be tuned to below 0.3 eV by doping with As. Black
phosphorus-arsenic (bPAs) has been demonstrated to achieve light detection from UV to THz.

Despite the very high responsivity reported in hybrid graphene-based
phototransistors, their power consumption, electronic readout circuits,
and noise are all determined by the zero-bandgap of graphene, which
leads to a large dark current. Alternative 2D materials, especially TMDs,

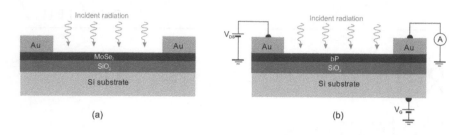

FIGURE 7.1   Cross-section view of 2D material photodetectors: (a) MoSe$_2$ photoconductor and (b) bP-based FET.

have been considered as potential replacements for graphene for transistor channels. The use of 2D TMD channels with a bandgap of 1–2 eV is of particular promise, offering lower leakage current during the operation of the transistor in the depletion mode.

As with graphene-based photodetectors, ultrahigh responsivity and ultrashort time response cannot be obtained at the same time in practice in 2D-related materials. In 2D materials and their hybrid systems, the photogating effect can be realized in two ways, as is described in Section 6.1.2, namely by carriers trapping in the localized states or (in hybrid photodetectors) by carriers trapping and transferring one type of carrier to the 2D materials.

In estimation of detectivity in layered detectors, the shot noise from the dark current is often considered to be the main source of noise. However, for photodetectors with a high photogain and operating at room temperature, the other two noise sources, thermal noise and generation-recombination (g-r) noise, cannot be totally neglected. In the case of devices with a high photogain and a low response speed induced by a long carrier lifetime, the light-induced g-r noise must be considered. In addition, low-frequency noise (1/$f$) should be investigated for various 2D photodetectors, because it is considered to be an important metric for evaluating the performance of these detectors.

Examples of the performance of infrared photodetectors, based on 2D materials and their hybrid structures, including photodetectors with cutoff wavelength above 1-μm, are collated in Table 7.1 [5,10–15]. Topological insulators can be promising candidate materials for broadband photodetection. Metallic surface state can result in a strong optical absorption, which has been demonstrated for the Bi$_2$Te$_3$/Si heterojunction. The spectral responsivity of this device covers a broadband response from 370.6 nm to 118 μm [15].

TABLE 7.1 Infrared Photoresponsivity of 2D Materials at Room Temperature

| Material | Wavelength (μm) | Responsivity (A/W) | Bias (V) | Gain (%) | Time | Peak Detectivity (Jones) | Ref. |
|---|---|---|---|---|---|---|---|
| MoS$_2$ | 0.5–1.1 | 0.1 | −10 | 25% | < 15 ms | | [5] |
| WS$_2$ (CVD) | 0.5–0.9 | $3.5 \times 10^5$ | 2 | $1 \times 10^5$ | 23 ms | $10^{14}$ | [5] |
| In$_2$Se$_3$ | 0.4–0.94 | $9.8 \times 10^4$ | 0.05 | | 9 s | $3.3 \times 10^{13}$ | [5] |
| GeAs | 1.6 | 6 | | | 3 s | | [5] |
| MoS$_2$/PbS | 0.4–1.5 | $6 \times 10^5$ | 1 | | 0.35 s | $7 \times 10^{14}$ | [5] |
| MoS$_2$/Si | 0.4–1.0 | 0.9082 | −2 | | 56 ns; 825 ns | $1.889 \times 10^{13}$ | [5] |
| MoS$_2$/bP | 0.5–1.6 | 22.3 | 3 | 50 | 15 μs; 70 μs | $3.1 \times 10^{13}$ | [5] |
| MoS$_2$/G/WS$_2$ | 0.4–2.4 | $1 \times 10^4$ | 1 | $10^6$ | 53.6 μs; 30.3 μs | $1 \times 10^{15}$ | [5] |
| bP (gated-photocon.) | 3.5 | 10 | 0.5 | 270 | | $6 \times 10^{10}$ | [10] |
| bP/MoS$_2$ (p-n hetero) | 4.3 | | 0.5 | | ~1 ms | $2 \times 10^9$ | [11] |
| bAsP (phototransistor) | 2–8 | $(30\text{–}10) \times 10^{-3}$ | 0 | | | $3 \times 10^8$ | [11] |
| PtSe$_2$ (phototransistor) | 0.6–10 | 4.5 | | | 1.1, 1.2 ms | $7 \times 10^8$ | [12] |
| PdSe$_2$ (phototransistor) | 1–10.6 | ~45 | 1 | $10^3$–49 | 74.5, 93.1 ms | $1 \times 10^9$ | [13] |
| PdSe$_2$/MoS$_2$ (p-n hetero) | 1–10.6 | ~4 | 1 | | 65.3, 62.4 μs | $8 \times 10^9$ | [13] |
| Gr/Ti$_2$O$_3$ | 10 | 300 | 0.1 | | 1.2, 2.6 ms | $7 \times 10^8$ | [14] |
| Bi$_2$Te$_3$/Si | UV-THz | 1 | −5 | | 0.1 s | $2.5 \times 10^{11}$ (635 nm) | [15] |

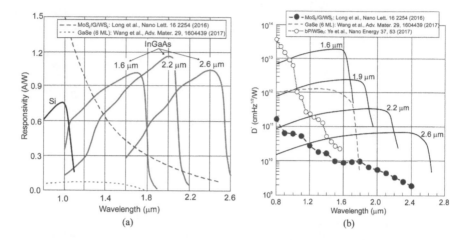

FIGURE 7.2 Comparison of (a) spectral responsivity and (b) detectivity of 2-D material photodetectors in SWIR spectral range with silicon and InGaAs photodiodes operating at room temperature.

Figure 7.2 compares the spectral responsivity and detectivity of representative 2D-material photodetectors in the SWIR spectral range with silicon and InGaAs photodiodes operated at room temperature. Generally, the responsivity and detectivity of layered material photodetectors are lower in comparison with the most popular commercial InGaAs photodiodes with a cutoff wavelength range below 3-μm.

## 7.2 MIDDLE- AND LONG-WAVELENGTH INFRARED DETECTORS

As shown in Table 7.1, in the longer wavelength infrared spectral region above 3-μm, black phosphorus (bP), black phosphorus-arsenic $As_xP_{1-x}$ (bPAs) alloys, and layered semiconductors with narrow bandgaps and high mobilities among the transition metal dichalcogenides (TMDs) are of great significance toward the realization of high-performance 2D infrared detectors.

### 7.2.1 Black Phosphorus Photodetectors

The attention paid to 2D bP, rediscovered since 2014, is as described in Section 5.2.2. bPAs as a monolayer or as a few-layer material, has shown attractive properties, such as high mobility, anisotropic optical properties, and a highly tunable thickness-dependent direct bandgap, spanning from 0.3 eV in the bulk case to 1.7 eV in the monolayer case. In this context, black phosphorus is evaluated as a natural compromise between graphene and TMDs.

The first bP paper of great significance, which revealed the potency of bP for practical infrared applications, was published by Guo *et al.* in 2016 [16]. The authors described a bP photoconductor with room-temperature responsivity of 82 A/W at a wavelength of 3.39 µm. This high responsivity is attributed to the photogating effect and the shallow long trap-level-induced carrier lifetime. Usually, a high photogain limits the bandwidth. For a light chopping frequency greater than 10 kHz, the responsivity value of 60 mA/W results from the intrinsic photovoltaic effect.

Bullock *et al.* [17] have demonstrated a two-terminal bP/MoS$_2$ MWIR heterojunction photodiode (Fig. 7.3) that achieved external quantum efficiencies of 35% and detectivities as high as $1.1 \times 10^{10}$ cmHz$^{1/2}$/W at $\lambda = 3.8$ µm at room temperature. These values are similar to those achieved by the current commercially available state-of-the-art room-temperature photodiodes (Figs. 3.15 and 3.16). Furthermore, by leveraging the anisotropic optical properties of bP, the first bias-selectable polarization-resolved photodetector was demonstrated.

The heterojunction photodiode consists of a bP/MoS$_2$ heterojunction in which a thin (~10–20 nm) n-type MoS$_2$ layer acts as an electron contact and an MWIR window. Holes are contacted *via* a rear back-reflector Au pad. As is shown in the band diagram of Fig. 7.3(b), the MoS$_2$ heterojunction provides asymmetric band offsets with the bP, allowing the flow of electrons to the MoS$_2$ contact, while blocking the flow of holes.

The polarized-resolved device uses two bP/MoS$_2$ heterojunctions, with orthogonal bP crystal orientations separated by a common MoS$_2$ electron contact. Each bP layer has an isolated hole contact. This configuration is similar to the two-color back-to-back photodiodes that had previously

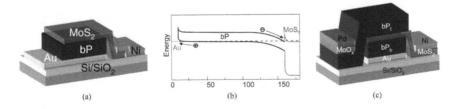

FIGURE 7.3 Two-terminal bP/MoS$_2$ MWIR heterojunction photodiodes: (a) schematic of device configuration, (b) energy band diagram of the device (a) under equilibrium, and (c) schematic of polarization-resolved bP/MoS$_2$ heterojunction photodiode (after Ref. [17]).

been demonstrated in two-color HgCdTe and III-V semiconductor photo-diodes [18]. The top $MoO_x$/Pd stack is an effective hole contact because of its large work function.

By varying the composition of arsenic, x, in the $As_xP_{1-x}$, the bandgap correspondingly changes from 0.3 to 0.15 eV. This change in the energy gap suggests that b-AsP may interact with light, where the wavelength is up to 8.3 μm. Long *et al.* [11] reported $bAs_{0.83}P_{0.17}$ long-wavelength IR pho-todetectors, with room-temperature operation to 8.2 μm. The design of the phototransistor is similar to that shown in Fig. 7.1(b). The b-AsP thin film flakes of b-AsP, ranging from 5- to 20-nm thick, were mechanically exfoliated from bulk b-AsP samples onto a highly doped silicon substrate covered by 300-nm $SiO_2$ in a glove box. After device fabrication by stan-dard electron-beam lithography, metallization, and a lift-off process, the spin-coated thin layer of polymethyl methacrylate (PMMA) was depos-ited to protect the samples from oxidation in the air.

Figure 7.4 schematically shows the photovoltaic response of b-AsP devices in which photogenerated electron-hole pairs are separated at the b-AsP/metal junctions. The photocurrent is mainly generated at the reverse-biased b-AsP/drain junction [Fig. 7.4(a)] if the channel is p-type doped. In the case of a slightly n-type doped b-AsP, the photocurrent is mainly generated at the reverse-biased b-AsP/source junction [Fig 7.4(b)]. The photocurrent has opposite polarity at the two contacts, due to the opposite junction bias direction. Detailed studies of the photoresponse mechanism revealed a more complex phenomenon, with contributions

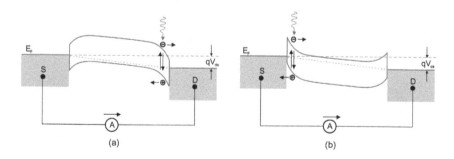

(a)  (b)

FIGURE 7.4 Schematic diagrams of energy band structure for MWIR b-AsP photovoltaic detector with different doping types under a bias voltage $V_{ds}$: (a) the sample of b-AsP working at the p-type region and (b) the device working at the n-type region. The black horizontal arrows indicate the direction of the photo-current (after Ref. [11]).

from the photogating and PFE effects. The photocurrent is negative, relative to $V_{ds}$, and shows very weak gate-dependence in the highly doped regime, with dominant contributions of photothermoelectric (PTE) and bolometric effects.

A well-known strategy to suppress dark current involves using heterojunctions, in this case, using 2D van der Waals (vdW) heterojunctions. Integrability is an inherent trait of 2D materials, by which different 2D flakes can be sequentially stacked into vdW heterojunctions. Following this idea, Long et al. [11] fabricated photodetectors, based on a b-AsP/MoS$_2$ heterostructure, with room-temperature detectivity of about $5 \times 10^9$ cmHz$^{1/2}$/W at 4.3 μm. Figure 3.15 compares spectral detectivity curves of two types of bAs$_{0.83}$P$_{0.17}$ photodetectors (namely, phototransistor and vdW heterojunction) which were comparable with commercially available ones. More recently, Tan et al. [19] have demonstrated that doping bP with carbon, to form bPC, can extend the detectable wavelength to 8 μm, and a peak responsivity of 2163 A/W was achieved from a bPC-based phototransistor at 2 μm.

Amani et al. [10] studied the response spectrum and responsivity of bP and bPAs alloys (composition range 0–91% of As) as the active regions of phototransistors. Figure 3.15 shows the spectral dependence of detectivity for b-PAs gated-photoconductors (phototransistors), with the cutoff wavelength tuned from 3.9 μm to 4.6 μm. It is shown that, for optimized devices with a thickness of ~28 nm, the peak detectivity is $6 \times 10^{10}$ cmHz$^{1/2}$/W for bP and $2.4 \times 10^{10}$ cmHz$^{1/2}$/W for bP$_{0.91}$As$_{0.09}$ at room temperature, with a bandwidth of 117 kHz [10]. This detectivity is a record achievement for bPAs 2D-material photodetectors in the MWIR spectral range, and is almost one order of magnitude greater than that from commercially available HgCdTe photodiodes.

Figure 7.5 compares the peak detectivity of HgCdTe photodiodes [20] and InAs/GaSb type-II superlattice IB QCIPs [21], operating at room temperature with bPAs and noble metal dichalcogenide photodetectors. We can see that, in the MWIR spectral range, the performance of bPAs detectors outperformed commercially available uncooled HgCdTe photodiodes; in LWIR, however, the detectivity of noble transition metal dichalcogenide photodetectors is the highest. Due to strong covalent bonding of III-V semiconductors, IB QCIPs can be operated at temperatures of up to 400°C, which is not capable of being achieved by their HgCdTe counterpart, especially b-PAs, due to their fatal problem of structural instability. Furthermore, the response time of HgCdTe and IB QCIPs detectors

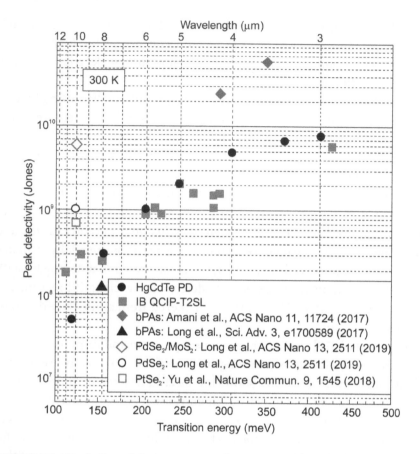

**FIGURE 7.5** Peak detectivity comparison between HgCdTe photodiodes, InAs/GaSb type-II superlattice IB QCIPs, and representative 2D-material photodetectors operating at room temperatures. Experimental data for HgCdTe photodiodes are taken according to the Vigo catalogue [20]. Data for cascade infrared detectors are taken from many papers, including Ref. [21].

operated at room temperature, typically of the order of nanoseconds, is considerably shorter than that of b-PAs photodetectors.

## 7.2.2 Noble Transition Metal Dichalcogenide Photodetectors

Narrow-bandgap 2D noble metal dichalcogenides could be novel platforms for room-temperature LWIR photodetectors. Theoretical simulations predict that group X transition-metal dichalcogenides (TMDs) (Ni, Pd, Pt) are promising narrow-bandgap semiconductors, with ~0–0.25 eV [22–25]. The room-temperature mobility of bulk materials is greater than

1000 cm$^2$/Vs. For 2D materials, carrier mobilities greater than 200 cm$^2$/Vs and air-stable properties have been demonstrated in recent years [22,26].

The first few-layered PtS$_2$ phototransistor on a hBN substrate, operating within the visible range, was demonstrated in 2017 [27]. A high photoresponsivity, up to $1.56 \times 10^3$ A/W, and an ultrahigh photogain, $2 \times 10^6$, was achieved, benefiting from the short-channel and hole-trap states. Yu *et al.* reported an Ar-plasmon-treated bilayer PtSe$_2$, with a responsivity and response time of 4.5 A/W and 1 ms, respectively, at a wavelength of 10 μm [12]. The detectivity approaches $7 \times 10^8$ cmHz$^{1/2}$/W, which is higher than that of commercial bolometers operating in this wavelength range.

Black phosphorus photodetectors are challenging because of their instability in air. Investigations into air-stable group X TMD materials for LWIR applications have just started. A new step in their development has been provided by Long *et al.* in a recently published paper [13]. This paper is devoted to two types of PdSe$_2$ photodetectors, based on phototransistors and their heterostructure.

The PdSe$_2$ FETs were fabricated by a conventional electron-beam lithography process, using flakes with a thickness of 5–20 nm to ensure high carrier mobility and a relatively high absorption. Figure 7.6 shows the long-wavelength infrared photoresponse characteristics of a typical PdSe$_2$ phototransistor at room temperature. The responsivity decreases from 42.1 to 13.8 A/W at $V_{ds} = 1$ V under laser illumination of 10.6 μm, when the light power is increased from 1.42 to 5.67 nW. The photogating effect plays a crucial role in high photoconductive gain due to the long lifetime of combination induced by the trap states and the short carrier transit time. As shown in Fig. 7.6, the photogating effect decreases from 49 to 16 as the illumination power increases. This effect is particularly apparent when the responsivities of HgCdTe photodiodes, where $g = 1$, are compared with these of PdSe$_2$ phototransistors [Fig. 7.6(b)]. In the case of PdSe$_2$ phototransitors, the responsivity is about one order of magnitude higher than for HgCdTe photodiodes in the wide infrared spectral range. The photoresponse speed, one of the most important figures of merit, is rather slow. The rise/delay time, defined as the time required to transition from 10/90% of the stable photocurrent during the illumination on/off cycle, is 74.5 ms/93.1 ms under 10.6-μm laser illumination. The response time under shorter-wavelength light is much faster, by about two orders of magnitude.

To decrease the dark current density, a heterojunction device is usually employed. The p-n heterojunction could be fabricated by stacking p-type

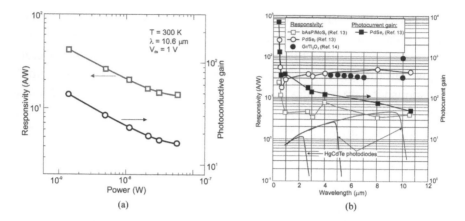

FIGURE 7.6   Long-wavelength infrared photoresponse of a typical PdSe₂ photo-transistor at room temperature: (a) power dependence responsivity and gain at $V_{ds} = 1$ V, (b) responsivity and photocurrent gain as a function of wavelength at 1 V bias voltage (after Ref. [13]).

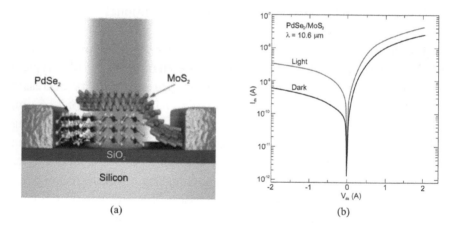

FIGURE 7.7   PdSe₂/MoS₂ heterostructure device: (a) schematic structure of the heterojunction, (b) $I_{ds}(V_{ds})$ dark and illumination characteristics at room temperature (after Ref. [13]).

PdSe₂ with other n-type 2D materials. Long *et al.* [13] selected n-type MoS₂, deposited on p-type PdSe₂, to form vdW heterojunction. A schematic structure of this device is shown in Fig. 7.7(a). Standard electron-beam evaporation was used for fabrication of the metal electrodes (5-nm Ti and 50 nm-Au).

Figure 7.7(b) shows a satisfactory device rectification effect of the output curve under dark conditions. The rectification ratio is typically about 100 at a 2 V bias, which indicates that a built-in electrical field exists at the interface. This is characterized by a broadband photoresponse from 450 nm to 10.6 μm, which is shown in Fig. 7.6(b). In the spectral range between 1 and 10.6 μm, the responsivity becomes stable at ~4 A/W.

The current noise of the $PdSe_2/MoS_2$ heterostructure is significantly suppressed, relative to that of the $PdSe_2$ field-effect transistor (FET). At the low-frequency range, $1/f$ noise dominates the noise current contribution. It is well known that the low-frequency flicker noise originates from the fluctuation of carriers being trapped and detrapped by defects and disorder and exists widely in 2D materials [28]. If the frequency increases beyond 1 kHz, the noise current of the $PdSe_2/MoS_2$ heterostructure decreases to the Johnson noise level, whereas, for the $PdSe_2$ FET devices, the current noise is three orders of magnitude higher than the Johnson noise level. This result indicates that the built-in electric field at the heterojunction effectively depresses the noise level, an effect which is highly desirable. In consequence, in spite of the higher responsivity of $PdSe_2$ FET, in comparison with the heterostructure [Fig. 7.6(b)], the detectivity of the $PdSe_2/MoS_2$ heterojunction is higher, at $8 \times 10^9$ cmHz$^{1/2}$/W at $\lambda = 10.6$ m and $T = 300$ K (Figs. 3.15 and 7.5). This is a record achievement, because this value of detectivity is more than one order of magnitude larger than that of commercial HgCdTe photodiodes.

## 7.3 TERAHERTZ DETECTORS

Diverse optoelectronic properties of 2D materials cover a wide range of the electromagnetic spectrum. Recently, efforts have increased to also develop the proof-of-concept for THz photodetectors. The finite and direct bandgap in both the bulk ($E_g \approx 0.35$ eV) and monolayer bP ($E_g \approx 2$ eV) phase, and its considerably greater mobility (>1000 cm$^2$/Vs), make it a good trade-off between graphene and TMDs. Furthermore, the achievable on-off transistor ratio greater than $10^5$ makes bP a suitable candidate for detection of THz-frequency light.

In 2015, Viti *et al.* first demonstrated that bP THz photodetectors operated at room temperature [29]. They exploited the integration of a mechanically $SiO_2$-encapsulated bP flake in an antenna-coupled top-gate FET, as is shown in Fig. 7.8. A standard adhesive tape technique was used to transfer the flake onto a 300-nm thick $SiO_2$ layer on the top of a 300-μm intrinsic silicon wafer. The selected bP flakes, with thickness of about 10 nm, coincide with the out-of-plane screening length and correspond to ≈16 layers of phosphorene. This choice ensures an ideal compromise between high mobility and large carrier density tunability.

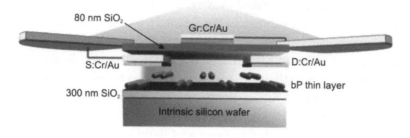

FIGURE 7.8    Vertical sketch of the bP-FET structure.

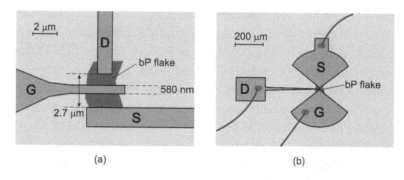

FIGURE 7.9    Fabrication of the bP-FET structure: (a) design of the device [the channel length ($L_c$) is 2.7 μm and the gate length ($L_g$) is 530 nm], (b) S and G electrodes are designed to form a 500-μm, 90° flare angle, planar bow-tie, antenna; the D electrode is connected to a rectangular bonding pad.

THz FET detectors have been fabricated by exploiting a combination of electron-beam lithography and metal evaporation. The source (S) and gate (G) electrodes were patterned in the shape of two halves of a planar bow-tie antenna, having a total length $2L = 500$ μm, and a flare angle of 90°, in resonance with the 0.3 THz radiation. Figure 7.9 shows the device layout [29].

The photodetection mechanism in bP-based THz FETs was identified as a consequence of three mechanisms, including photothermoelectric, bolometric, and plasma-wave rectification effects [29], and the anisotropy of bP [30]. For example, Fig. 7.10 shows the extrapolated *NEP* curves for the above three mechanisms, the minimum of which reaches 7 nW/Hz$^{1/2}$, 10 nW/Hz$^{1/2}$, and 45 nW/Hz$^{1/2}$ for the bP-bolometer, plasma-wave, and thermoelectric detector, respectively. The responsivity value of ~5–8 A/W at about 0.3 THz permits the use of the bP FET device in application for real-time pharmaceutical and quality control imaging [30].

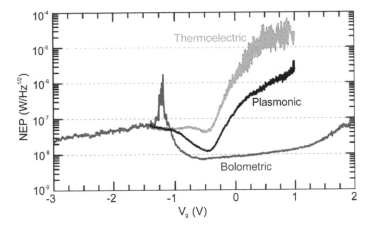

FIGURE 7.10 Noise equivalent power as a function of gate voltage ($V_g$) for the plasma-wave sample (0.29 THz), thermoelectric sample (0.32 THz) and bolometric sample detectors (after Ref. [30]).

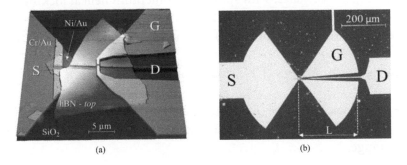

FIGURE 7.11 hBN/bP/hBN heterostructure THz FET detectors: (a) AFM tomographic image of the top-gate FET, (b) optical microscopy image of the fabricated device. The bow arm length is $L = 250$ μm (after Ref. [31]).

To prevent fast degradation of the exfoliated bP flake at ambient temperature, Viti *et al.* embedded a bP flake within a natural semiconductor heterostructure, formed by multilayered h-BN crystals, to devise hBN/bP/hBN heterostructure THz FET detectors, as shown in Fig. 7.11. The heterostructure was fabricated as previously (see Fig. 7.8) over a high-resistivity silicon substrate covered with a $SiO_2$ cap layer. A 40-nm thick hBN flake with lateral dimensions $20 \times 20$ μm² was put on the $SiO_2$/Si substrate to facilitate the subsequent alignment step, transferring the bP flake onto the lower hBN layer. Before the deposition of the hBN top layer, the source (S) and drain (D) contacts were defined through a combination of electron-beam lithography and metal deposition on the bP flake.

The FET hBN/bP/hBN heterostructure, operating at 295 GHz, is characterized by a minimum *NEP* of 100 pW/Hz$^{1/2}$ and maximum $R_v = 38$ V/W at 4 K, whereas, at room temperature, the corresponding values are ~ 10 nW/Hz$^{1/2}$ and ~2 V/W, respectively. hBN as an atomic dielectric constitutes a good encapsulating material for bP. In addition, hBN is a valuable gate dielectric, which, in turn, allows high gate-to-channel capacitance for tuning the device transport and optical properties. Viti *et al.* [32] reviewed recent research on bP detectors operating in a wider spectral range, from 0.26 THz to 3.4 THz, with emphasis on the future challenges in the fabrication of stable and reliable optical and electronic technologies.

Recently, it was demonstrated that topological insulators can also be promising candidate materials for broadband photodetection, including the THz range. Topological insulators (TIs) represent a novel quantum phase of matter, characterized by a semiconducting bulk and topologically protected surface states, with spin and momentum helical locking and a Dirac-like band structure [33,34]. 2D TIs are associated with gapless edge states, and three-dimensional (3D) insulators with gapless topological surface states (TSS) [35].

Intriguing properties of TIs result from:

- The gapless state of TSS, which enable carrier generation by light absorption over a very wide energy spectrum, including the THz range,

- Optoelectronic properties, which can be engineered by material stoichiometry [36],

- High TSS mobility, even higher than that of graphene, due to the topological protection that prevents back-scattering effects [37], and

- 2D TSS electron gas of 3D TIs supports a collective excitation (Dirac plasmons) in the THz range [38].

Most of the published papers concern visible and short-wavelength infrared TIs photodetectors [39]. The promising application of THz plasmonics with TIs is related to the rectification of THz radiation *via* the excitation of plasma waves in the active channel of antenna-coupled FETs.

The first demonstration of THz detection, mediated by TSS in top-gated nanometer FETs, exploiting thin $Bi_2Te_{3-x}Se_x$ flakes, was demonstrated by

Viti *et al.* in 2016 [40] (Fig. 7.12). Flakes having a total length < 2 µm have been selected as active nonlinear elements of FETs, in order to reduce both the parasitic capacitance and the resistances in the ungated transistor regions. By engineering the $Bi_2Te_{3-x}Se_x$ stoichiometry, and mediating the room temperature THz detection with overdamped plasma-wave oscillations, a maximum responsivity of 3.0 V/W, and a minimum *NEP* of ~10 nW/Hz$^{1/2}$ have been achieved for a 292.7-GHz impinging radiation frequency. Using this detector, the 400 × 700 pixel image of a target object was acquired with a time constant of 20 ms. This experiment gives a realistic account of the exploitation of TSS for large-area, fast THz imaging.

Another structure of TI THz photodetectors has also been proposed. Yao *et al.* [41] proposed a $Bi_2Te_3$-Si heterojunction detector, schematically shown in Fig. 7.13(a). The $Bi_2Te_3$ film, 300-µm in diameter and 100-nm in thickness, was grown onto the Si wafer by pulsed laser deposition. Pt and Ag electrode contacts were then deposited on the top surface of the $Bi_2Te_3$ and the bottom surface of the Si, respectively. The device demonstrated photoresponse at room temperature in the wavelength range from the ultraviolet (370.6 nm) to terahertz (118 µm).

An innovative strategy for the direct detection of THz photons at room temperature, based on a subwavelength metal- $Bi_2Se_3$-metal structure, has been proposed in Ref. [42] [Fig. 7.13(b)]. In this device, the contact metal converts incident radiation into a localized surface-plasmon field, which drives the TSS back and forth. The measured room-temperature responsivity of the device, working in the self-powered and bias modes at

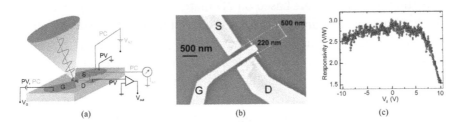

(a)          (b)          (c)

FIGURE 7.12 Top-gated nanometer FET exploiting a thin $Bi_2Te_{3-x}Se_x$ flake: (a) schematics of the THz detection principle, where photovoltage and photocurrent modes are depicted in black and purple, respectively; (b) false-color SEM image of the top-gated FET; S, G, and D stand for source, gate, and drain, respectively; and (c) experimental responsivity measured at room temperature at a fixed frequency $\nu = 292.7$ GHz, while sweeping the gate-bias $V_g$ in the range (−10 V, + 10 V) (after Ref. [40]).

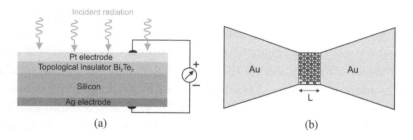

(a)          (b)

FIGURE 7.13   Structures of THz photodetectors with topological insulators: (a) $Bi_2Te_3$-Si heterojunction in a two-pole structure (the forward bias is defined as applying a positive bias on the $Bi_2Te_3$ side), (b) metal-$Bi_2Se_3$-metal structure (the spacing distance $L$ between the two gold contacts is much shorter than the wavelength radiation).

0.3 THz, was 75 and 475 A/W, respectively. The *NEP* value of $3.6 \times 10^{-13}$ W/Hz$^{1/2}$ and the detectivity of $2.17 \times 10^{11}$ cmHz$^{1/2}$/W was achieved for a device working under a bias current of 50 mV. These results are crucial for the successful use of TIs in the field of THz imaging applications.

## 7.4 2D-MATERIAL DETECTOR PERFORMANCE – THE PRESENT STATUS

Development of pristine 2D-material high-sensitivity photodetectors is determined by two major challenges: the low optical absorption in thin active regions (~ 100–200 nm) and the short photocarrier lifetime. As with graphene-based detectors, the photodetectors are limited by trade-offs between high responsivity, ultrafast response time, and broadband operation. Photodetectors based on 2D materials display a large variation in their current responsivity and response time [2,43], covering about nine orders of magnitudes, as is shown in Fig. 7.14. The current studies are mostly limited to exfoliated materials from bulk-layered crystals, with very limited yields, reproducibility, and scalability. In consequence, the divergence in the responsivity values is large. Moreover, the photopaging effect has a strong influence on the relationship between current responsivity and response time, which is shown in Fig. 7.14(b).

To enhance the infrared absorption, multiple layers, instead of a single layer, are selected. High photogain is often realized by using the 2D material as the fast transfer channel for charge carriers. However, as mentioned in Chapter 6, their general drawback is a very slow response time attributed to traps and enhanced capacitance. The response time at room temperature is typically longer than ~0.1 ms, which indicates a considerably

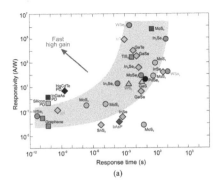 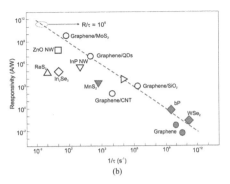

(a)          (b)

FIGURE 7.14  Room-temperature current responsivity against response time for 2D-material photodetectors: (a) in comparison with commercial silicon, InGaAs, and HgCdTe photodiodes (many data are taken from Ref. [43]), and (b) for part low-dimensional photodetectors (after Ref. [44]); the dashed line represents a typical magnitude order of gain-bandwidth product for traditional high-performance thin-film photodetectors, and the white symbol indicates that this is a photogating-enhanced photodetector.

longer response time in comparison with commercial silicon, InGaAs, or HgCdTe photodiodes; for HOT photodiodes, the response time is typically between 1 and 10 ns. The upper-left blank panel of Fig. 7.14(a) indicates the lack of photodetectors with both ultrahigh responsivity and ultrafast response speed. This observation is also supported by Fig. 7.14(b). It is also well known that the high gain induced by a long carrier lifetime can also be caused by both a large generation-recombination noise and a slow speed, which are incompatible with improving the device detectivity. In detector focal plane arrays, the suitable response time depends on the pixel number, signal readout mode, driving integrated circuit design rules, and other factors.

Figure 7.15 summarizes the photoresponsivity and response time of different 2D-material photodetectors. It is shown that black phosphorus (bP) falls into a region between graphene and TMDs. However, degradation in air and other environments is an unresolved issue that may limit future applications of bP [45]. bP degrades rapidly under ambient conditions, affecting its structure and properties. In particular, the role of different ambient species has remained controversial. Considerably better stability is observed for noble transition metal dichalcogenides, with record room-temperature detectivity measured for $PdSe_2/MoS_2$ heterostructures of about $10^{10}$ Jones, in the long-wavelength infrared range [13].

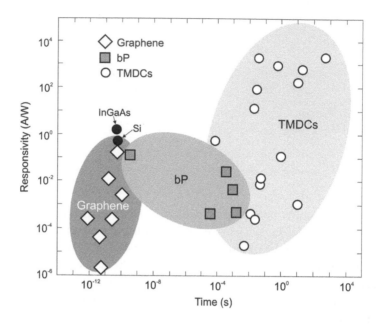

FIGURE 7.15  Summary of the responsivity and response time of photodetectors based on different groups of 2D materials: graphene, black phosphorus (bP), and transition metal dichalcogenides (TMDs).

Another study on the 2D material gives the experimentally measured relationship between detectivity and response time, as shown in Fig. 7.16, for photodetectors operated mainly in the visible and SWIR spectral regions [46]. The large variation in responsivity and response time (Fig. 7.14) results in a change in the $D^*f$ product of about ten orders of magnitude, between $10^7$ and $10^{17}$ Jones·Hz.

Although photodetection platforms based on 2D materials have demonstrated a variety of potential applications, outstanding challenges remain to be addressed in order to exploit the distinct advantages of these new materials. The prospects for commercialization will depend not only on the detector performance, but also on the distinct advantages in the ability to fabricate large-scale high-quality 2D materials at a low cost. The final goal is to establish large-scale integration of 2D crystals with existing photonic and electronic platforms, such as CMOS technologies.

Most of the high-performance 2D-material photodetectors, consisting of thin layers or heterostructures, are achieved by mechanical exfoliation. There is still a significant need for improvement of 2D-material technology in order to obtain well-developed devices of industrial applicability. One of the critically important issues is the stability and quality of 2D

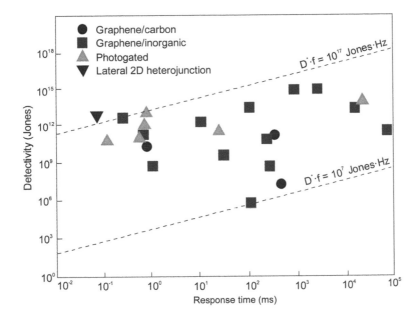

FIGURE 7.16  The state-of-the-art graphene heterojunction devices (after Ref. [46]).

materials. Although many efforts have been devoted to synthesizing large-sized single-crystal 2D materials, it is still hard to obtain single-crystals larger than millimeter size, and the grain size and sample size of TMDs are even smaller. Using the CVD-grown technique, relatively large size samples can be obtained, but they have defects and contaminations, which can cause sample degradation over time, resulting in devices which are not stable.

Generally, photodetectors made from TMD-layered semiconducting material operate at the visible and near-infrared regions, and generally their high sensitivity does not coincide with a fast response time. In comparison with a response time measured in the nanosecond range for HgCdTe photodiodes and type-II superlattice IB QCIPs operating at room temperature, the response time of TMD is typically in the millisecond range. This indicates that 2D-infrared photodetectors operating at room temperature are not ready for real-life applications. The performances of black phosphorus-arsenic alloy detectors are comparable to commercial HgCdTe photodiodes. However, the instability of the surface, due to chemical degradation under ambient conditions, remains a major impediment to its prospective applications. More promising are stable noble TMD photodetectors like $PdSe_2/MoS_2$ heterojunctions, with record detectivity in the LWIR range at room temperature. However, their practical application lies in perfect material synthesis and processing. To achieve their potential, 2D-material photodetectors still have a long way to go.

## REFERENCES

1. F.H.L. Koppens, T. Mueller, Ph. Avouris, A.C. Ferrari, M.S. Vitiello, and M. Polini, "Photodetectors based on graphene, other two-dimensional materials and hybrid systems", *Nature Nanotechnology* **9**, 780–793 (2014).
2. M. Buscema, J.O. Island, D.J. Groenendijk, S.I. Blanter, G.A. Steele, H.S.J. van der Zant, and A. Castellanos-Gomez, "Photocurrent generation with two-dimensional van der Waals semiconductor", *Chemical Society Reviews* **44**, 3691–3718 (2015).
3. X. Li, L. Tao, Z. Chen, H. Fang, X. Li, X. Wang, J.-B. Xu, and H. Zhu, "Graphene and related two-dimensional materials: Structure-property relationships for electronics and optoelectronics", *Applied Physics Reviews* **4**, 021306–1–31 (2017).
4. G. Wang, Y. Zhang, C. You, B. Liu, Y. Yang, H. Li, A. Cui, D. Liu, and H. Yan, "Two dimensional materials based photodetectors", *Infrared Physics and Technology* **88**, 149–173 (2018).
5. F. Wang, Z. Wang, L. Yin, R. Cheng, J. Wang, Y. Wen, T.A. Shifa, F. Wang, Y. Zhang, X. Zhan, and J. He, "2D library beyond graphene and transition metal dichalcogenides: a focus on photodetection", *Chemical Society Reviews* **47**(16), 6296–6341 (2018).
6. M. Long, P. Wang, H. Fang, and W. Hu, "Progress, challenges, and opportunities for 2D material based photodetectors", *Advanced Functional Materials* **29**, 1803807 (2018).
7. Y. Wang, W. Wu, and Z. Zhao, "Recent progress and remaining challenges of 2D material-based terahertz detectors", *Infrared Physics and Technology* **102**, 103024 (2019).
8. P. Wang, H. Xia, Q. Li, F. Wang, L. Zhang, T. Li, P. Martyniuk, A. Rogalski, and W. Hu, "Sensing infrared photons at room temperature: from bulk materials to atomic layers", *Small* **46**(13), 1904396 (2019).
9. J. Cheng, C. Wang, X. Zou, and L. Liao, "Recent advances in optoelectronic devices based on 2D materials and their heterostructures", *Advanced Optical Materials* **7**, 1800441 (2019).
10. M. Amani, E. Regan, J. Bullock, G.H. Ahn, and A. Javey, "Mid-wave infrared photoconductors based on black phosphorus-arsenic alloys", *ACS Nano* **11**, 11724–11731 (2017).
11. M. Long, A. Gao, P. Wang, H. Xia, C. Ott, C. Pan, Y. Fu, E. Liu, X. Chen, W. Lu, T. Nilges, J. Xu, X. Wang, W. Hu, and F. Miao, "Room temperature high-detectivity mid-infrared photodetectors based on black arsenic phosphorus", *Science Advances* **3**, e1700589 (2017).

12. X. Yu, P. Yu, D. Wu, B. Singh, Q. Zeng, H. Lin, W. Zhou, J. Lin, K. Suenaga, Z. Liu, and Q.J. Wang, "Atomically thin noble metal dichalcogenide: a broadband mid-infrared semiconductor", *Nature Communications* **9**, 1545 (2018).

13. M. Long, Y. Wang, P. Wang, X. Zhou, H. Xia, C. Luo, S. Huang, G. Zhang, H. Yan, Z. Fan, X. Wu, X. Chen, W. Lu, and W. Hu, "Palladium diselenide long-wavelength infrared photodetector with high sensitivity and stability", *ACS Nano* **13**, 2511–2519 (2019).

14. X. Yu, Y. Li, X. Hu, D. Zhang, Y. Tao, Z. Liu, Y. He, M.A. Haque, Z. Liu, T. Wu, and Q.J. Wang, "Narrow bandgap oxide nanoparticles coupled with graphene for high performance mid-infrared photodetection", *Nature Communications* **9**, 4299 (2018).

15. J. Yao, J. Shao, Y. Wang, Z. Zhao, and G. Yang, "Ultra-broadband and high response of the $Bi_2Te_3$–Si heterojunction and its application as a photodetector at room temperature in harsh working environments", *Nanoscale* **7**, 12535–12541 (2015).

16. Q. Guo, A. Pospischil, M. Bhuiyan, H. Jiang, H. Tian, D. Farmer, B. Deng, C. Li, S.-J. Han, H. Wang, Q. Xia, T.-P. Ma, T. Mueller, and F. Xia, "Black phosphorus mid-infrared photodetectors with high gain", *Nano Letters* **16**, 4648–4655 (2016).

17. J. Bullock, M. Amani, J. Cho, Y.-Z. Chen, G.H. Ahn, V. Adinolfi, V.R. Shrestha, Y. Gao, K.B. Crozier, Y.-L. Chueh and A. Javey, "Polarization-resolved black phosphorus/molybdenum disulfide mid-wave infrared photodiodes with high detectivity at room temperature", *Nature Photonics* **12**, 601–607 (2018).

18. A. Rogalski, *Infrared and Terahertz Detectors*, 3rd edition, CRC Press, Boca Raton, 2019.

19. W.C. Tan, L. Huang, R.J. Ng, L. Wang, D.M.N. Hasan, T.J. Duffin, K.S. Kumar, C.A. Nijhuis, C. Lee, and K.-W. Ang, "A black phosphorus carbide infrared phototransistor", *Advanced Materials* **30**, 1705039 (2018).

20. https://vigo.com.pl/wp-content/uploads/2017/06/VIGO-Catalogue.pdf

21. A. Rogalski, P. Martyniuk, and M. Kopytko, "Type-II superlattice photodetectors versus HgCdTe photodiodes", *Progress in Quantum Electronics* **68**, 100228 (2019).

22. A.D. Oyedele, S. Yang, L. Liang, A.A. Puretzky, K. Wang, J. Zhang, P. Yu, P.R. Pudasaini, A.W. Ghosh, Z. Liu, C.M. Rouleau, B.G. Sumpter, M.F. Chisholm, W. Zhou, P.D. Rack, D.B. Geohegan, and K. Xiao, "$PdSe_2$: Pentagonal two-dimensional layers with high air stability for electronics", *Journal of the American Society* **139**, 14090–14097 (2017).

23. X. Yu, P. Yu, D. Wu, B. Singh, Q. Zeng, H. Lin, Wu Zhou, J.Lin, K. Suenaga, Z. Liu, and Q.J. Wang, "Atomically thin noble metal dichalcogenide: a broadband mid-infrared semiconductor", *Nature Communications* **9**, 1545 (2018).

24. L. Pi, L. Li, K. Liu, Q. Zhang, H. Li, and T. Zhai, "Recent progress on 2D noble-transition-metal dichalcogenides", *Advanced Functional Materials* **29**, 1904932 (2019).

25. L.-H. Zeng, D. Wu, S.-H. Lin, C. Xie, H.-Yu Yuan, W. Lu, S.P. Lau, Y. Chai, L.-B. Luo, Z.-J. Li, and Y.H. Tsang, "Controlled synthesis of 2D palladium diselenide for sensitive photodetector applications", *Advanced Functional Materials* **29**, 1806878 (2019).

26. Y. Zhao, J. Qiao, Z. Yu, P. Yu, K. Xu, S.P. Lau, W. Zhou, Z. Liu, X. Wang, W. Ji, and Y. Chai, "High-electron-mobility and air-stable 2D layered PtSe$_2$ FETs", *Advanced Materials* **29**, 1604230 (2017).

27. L. Li, W. Wang, Y. Chai, H. Li, M. Tian, and T. Zhai, "Few-layered PtS$_2$ phototransistor on h-BN with high gain", *Advanced Functional Materials* **27**, 1701011 (2017).

28. A.A. Balandin, "Low-frequency 1/f noise in graphene devices", *Nature Nanotechnology* **8**, 549–555 (2013).

29. L. Viti, J. Hu, D. Coquillat, W. Knap, A. Tredicucci, A. Politano, and M.S. Vitiello, "Black phosphorus terahertz photodetectors", *Advanced Materials* **27**, 5567–5572 (2015).

30. L. Viti, J. Hu, D. Coquillat, A. Politano, W. Knap, and M.S. Vitiello, "Efficient terahertz detection in black-phosphorus nano-transistors with selective and controllable plasma-wave, bolometric and thermoelectric response", *Science Reports* **6**, 20474 (2016).

31. L. Viti, J. Hu, D. Coquillat, A. Politano, C. Consejo, W. Knap, and M.S. Vitiello, "Heterostructured hBN-BP-hBN nanodetectors at terahertz frequencies", *Advanced Materials* **28**, 7390–7396 (2016).

32. L. Viti, A. Politano, and M.S. Vitiello, "Black phosphorus nanodevices at terahertz frequencies: Photodetectors and future challenges", *APL Materials* **5**, 035602 (2017).

33. M.Z. Hasan and C.L. Kane, "Colloquium: Topological insulators", *Reviews of Modern Physics* **82**, 3045–3067 (2010).

34. X.-L. Qi and S.-C., Zhang, "Topological insulators and superconductors", *Reviews of Modern Physics* **83**, 1057–1110 (2011)

35. Y. Ando, "Topological insulator materials", *Journal of the Physical Society of Japan* **82**, 102001 (2013).

36. H. Zhang, C.-X. Liu, X.-L. Qi, X. Dai, Z. Fang, and S.-C. Zhang, "Topological insulators in Bi$_2$Se$_3$, Bi$_2$Te$_3$ and Sb$_2$Te$_3$ with a single Dirac cone on the surface", *Nature Physics* **5**, 438–442 (2009).

37. L. Fu, C.L. Kane, and E.J. Mele, "Topological insulators in three dimensions", *Physical Review Letters* **98**, 106803 (2007).

38. P. Di Pietro, M. Ortolani, O. Limaj, A. Di Gaspare, V. Giliberti, F. Giorgianni, M. Brahlek, N. Bansal, N. Koirala, S. Oh, P. Calvani, and S. Lupi, "Observation of Dirac plasmons in a topological insulator", *Nature Nanotechnology* **8**, 556–560 (2013).

39. W. Tian, W. Yu, J. Shi, and Y. Wang, "The property, preparation and application of topological insulators: A review", *Materials* **10**, 814 (2017).

40. L. Viti, D. Coquillat, A. Politano, K.A. Kokh, Z.S. Aliev, M.B. Babanly, O.E. Tereshchenko, W. Knap, E.V. Chulkov, and M.S. Vitiello, "Plasma-wave terahertz detection mediated by topological insulators surface states", *Nano Letters* **16**, 80–87 (2016).

41. J. Yao, J. Shao, Y. Wang, Z. Zhao, and G. Yang, "Ultra-broadband and high response of the $Bi_2Te_3$–Si heterojunction and its application as a photodetector at room temperature in harsh working environments", *Nanoscale* **7**, 12535–12541 (2015).

42. W. Tang, A. Politano, C. Guo, W. Guo, C. Liu, L. Wang, X. Chen, and W. Lu, "Ultrasensitive room-temperature terahertz direct detection based on a bismuth selenide topological insulator", *Advanced Functional Materials* **28**, 1801786 (2018).

43. J. Wang, H. Fang, X. Wang, X. Chen, W. Lu, and W. Hu, "Recent progress on localized field enhanced two-dimensional material photodetectors from ultraviolet-visible to infrared", *Small* **13**, 1700894 (2017).

44. H. Fang and W. Hu, "Photogating in low dimensional photodetectors", *Advanced Science* **4**, 1700323 (2017).

45. S. Kuriakose, T. Ahmed, S. Balendhran, V. Bansal, S. Sriram, M. Bhaskaran, and S. Walia, "Black phosphorus: ambient degradation and strategies for protection", *2D Materials* **5**(3), 032001 (2018).

46. D.-R. Chen, M. Hofmann, H.-M. Yao, S.-K. Chiu, S.-H. Chen, Y.-R. Luo, C.-C. Hsu, and Y.-P. Hsieh, "Lateral two-dimensional material heterojunction photodetectors with ultrahigh speed and detectivity", *ACS Applied Materials and Interfaces* **11**, 6384–6388 (2019).

# CHAPTER 8

# Colloidal Quantum Dot Infrared Detectors

THEORETICAL ESTIMATES CARRIED OUT by Martyniuk *et al.* [1] in 2008 indicate that the self-assembled quantum dot infrared photodetectors (QDIPs) are suitable for noncryogenic operation, especially in long-wavelength infrared (LWIR) regions. In practice, however, the reduced performance of QDIPs is the result of nonoptimal band structure and technological problems, such as QD size and density control. More recently, the attractive alternative to self-assembled epitaxial QDs has been colloidal quantum dots (CQDs), with better size tunability of the optical features and lower fabrication cost.

In the past decade, significant progress has been made in the fabrication of CQD photodetectors. Using this approach, an active region is constructed, based on 3D quantum-confined semiconductor nanoparticles, synthesized by inorganic chemistry. These nanoparticles could improve the performance of CQD photodetectors, compared with that of epitaxial QDs, as a result of many aspects, collated in Table 8.1 [2,3].

CQDs have attracted attention as a candidate material for a range of optoelectronic applications, including light-emitting diodes, lasers, optical modulators, solar cells, and photodetectors [2,3]. The first mass market for these nanoparticles appeared around 2015, as they started to be used as phosphors for TV displays [4].

TABLE 8.1    Advantages and Disadvantages of CQD Photodetectors in Comparison with Single-crystal QD Photodetectors

| Advantages | Disadvantages |
|---|---|
| • Control of dot synthesis and absorption spectrum by ability of QD size-filtering, which leads to highly uniform ensembles<br>• Much stronger absorption than in Stranski-Krastanov-grown QD, due to close-packing of CDs<br>• Considerable elimination of strains influencing the growth of epitaxial QDs by greater selection of active region materials<br>• Reduction of fabrication cost (using, e.g., such solutions as spin coating, inject printing, doctor blade, or roll-to-roll printing) compared to epitaxial growth<br>• Deposition methods are compatible with a variety of flexible substrates and sensing technologies, such as CMOS (e.g. direct coating on silicon electronics for imaging) | • Inferior chemical stability and electronic passivation of the nanomaterials in comparison with epitaxial materials<br>• Bipolar, inter-band (or excitonic) transitions across the CQD bandgap (e.g. electrons hopping among QDs and hole transport through the polymer), contrary to the intra-band transitions in the epitaxial QDs<br>• Insulating behavior due to slow electron transfer through many barrier interfaces in a nanomaterial<br>• Problems with long-term stability due to the large density of interfaces with atoms presenting different or weaker binding<br>• High level of $1/f$ noise due to disordered granular systems |

## 8.1  INTRODUCTION

CQDs operate in near/short-wave/mid-wave infrared (NIR/SWIR/MWIR) imaging (Fig. 8.1) and offer a promising alternative to the single-crystal IR materials [InGaAs, InSb, InAsSb, and HgCdTe, as well as type-II superlattices (T2SLs)].

CQDs belong to a wider class of low-dimensional materials, such as 2D materials, nanowires, and their hybrid structures. As noted in Chapters 6 and 7, effective active regions of photodetectors are fabricated by hybridization of 2D materials with CQDs. For this reason, this chapter provides a review of the fundamental properties of CQD photodetectors operating in the infrared spectral region.

Table 8.2 collates performance parameters of representative CQD photodetectors, whereas Fig. 8.2 plots representative values of responsivity for nanostructured materials *versus* their detection wavelengths [5]. As shown, the nanostructured materials are characterized by a broadband spectral response.

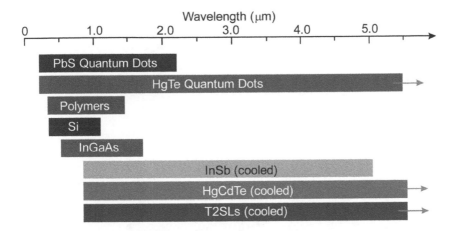

FIGURE 8.1   The wavelength ranges that can be detected by materials commonly used in imaging applications.

TABLE 8.2   The Colloidal Quantum Dots and Their Performance Parameters (after Ref. [5])

| Materials | Wavelength (μm) | Responsivity (A/W) | Detectivity (Jones) | Speed (Hz) | Pub. year |
|---|---|---|---|---|---|
| HgTe | 2–6.5 | – | – | – | 2018 |
| HgTe | 2.0 | – | $2 \times 10^{10}$ | 2 | 2017 |
| HgSe | 4.2–9 | 0.145 | – | – | 2017 |
| PbS | 1.3 | $1.0 \times 10^3$ | $1.8 \times 10^{13}$ | 18 | 2006 |

## 8.2  BRIEF VIEW

Typically, CQD photodetectors are fabricated using conducting-polymer/nanocrystal blends, or nanocomposites [6–10]. Nanocomposites often feature narrow-bandgap, II-VI (HgTe, HgSe) [11–13], PbSe or PbS [14–17] colloidal QDs. Usually, the reported IR photodetectors use CQDs embedded in conducting polymer matrices, such as poly[2-methoxy-5-(2-ethylhexyloxy)-1,4-phenylenevinylene] (MEH-PPV), and exhibit photodetection in the near-IR regime (1–3 μm), corresponding to the semiconductor nanocrystal bandgap energy [6].

Colloidal QD photodetectors typically comprise a single nanocomposite layer deposited on a glass slide by spin-casting, and large-area, two-terminal, vertical devices are fabricated using p-[indium-tin-oxide (ITO)] and n-type

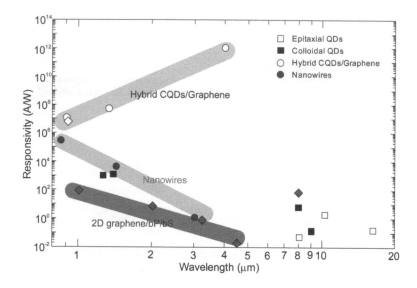

FIGURE 8.2   A summary of the spectral responsivity of infrared photodetectors based on different nanostructured materials (after Ref. [5]).

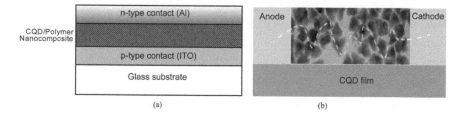

FIGURE 8.3   Colloidal quantum dot photodetector: (a) schematic diagram of device heterostructure in CQD/conducting polymer nanocomposites; (b) an SEM image of a PV QD detector with transport illustration of a photo-generated charge.

(aluminum) contacts, as shown schematically in Fig. 8.3(a). Figure 8.3(b) illustrates the capture and transport mechanism of a colloidal dot film.

The charge transport mechanisms in colloidal QD nanocomposites exhibit subtle differences, compared with epitaxial QDIPs. As shown in Fig. 8.4, the intra-band transitions are not exploited. Instead, bipolar, inter-band (or excitonic) transitions across the colloidal QD bandgap contribute to the photoresponse of the detector. In addition, since CQDs are electron acceptors and the polymers are typically hole conductors, the photogenerated excitons are dissociated at the QD/polymer interface.

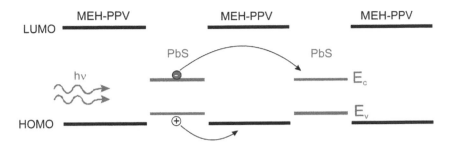

FIGURE 8.4   Schematic diagram of energy *vs.* position for inter-band transitions in PbS/MEH-PPV colloidal QD-conducting polymer nanocomposites, demonstrating photocurrent generation for IR photodetection (after Ref. [18]).

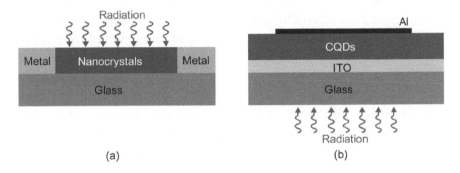

FIGURE 8.5   Two classes of CQD photodetectors: (a) the photoconductor, and (b) the photodiode.

Thus, photoconduction through the nanocomposite occurs as electrons hop among the QDs and holes transport through the polymer [18].

There are two main classes of CQD photodetectors: photoconductors and photodiodes (Fig. 8.5). A photoconductor typically consists of an active semiconducting material (single-crystalline, polycrystalline, amorphous) and two ohmic metal contacts. In the case of photodiodes, the depletion layer is formed between the nanocrystal film and an ohmic contact. A planar transparent ITO film forms the opposing ohmic contact. The photodiode is backside-illuminated through transparent ITO film.

Early studies of CQD-based detectors started from photoconductors. However, in spite of the simplicity of device architecture, the performance of CQD photoconductors was limited by dark current, $1/f$ noise, and difficulty in the precise control of doping concentration. Improved performance was achieved using photovoltaic devices, which, in principle, decreased $1/f$ noise and dark currents.

Although the performance of IR focal plane arrays (FPAs) remains paramount, especially in defense and security applications, the ability to produce FPAs at low cost is becoming an increasingly integral part of the implementation strategy for existing and emerging infrared imaging systems. It is expected that extension of the application of CQD-based devices will be significant, especially in the area of IR imaging, which is currently dominated by epitaxial semiconductor and hybrid technologies. Hybrid technology, due to the complexity of production stages, reduces yield and increases overall cost. The IR CQD-based photodetectors are an alternative solution without these limitations.

IR FPAs are usually fabricated from two separate wafers (Section 4.2), one bearing detector photodiodes and the other containing silicon read-out integrated circuits (ROICs), which are physically bonded (hybridized) together *via* indium bumps. The complexity of these multiple production steps (about 150 steps in total) reduces yield and increases the overall cost. The CQD-based vertical technology, shown in Fig. 8.5(b), permits fabrication of detector arrays directly onto ROIC substrates using a solution process [6,9], as illustrated in Fig. 4.14. Individual pixels are defined by the area of the metal pads arranged on the top of the ROIC surface. The 100-nm thickness of PbS CQD stacks can be used as an efficient absorption region [17]. The traditional hybrid FPAs are typically limited to small arrays (1-megapixel range), due to small detector wafers and low throughput. Pixel pitch, with a size of less than 10-μm, has already been demonstrated (Section 4.5). With a thin-film active layer integrated monolithically directly on top of the ROIC, submicron pixel size (i.e., below 1-μm) can be achieved (Fig. 8.6 [17]). At present, the state-of-the-art for CMOS image sensors is 0.9-μm [19].

Malinowski *et al.* [17,20] described a CMOS-compatible pixel stack, based on a PbS QD tunable absorption peak. As shown in Fig. 8.7(a), the absorption peak depends on the nanocrystal size. Using larger-sized crystals (5.5-nm diameter), the peak absorption occurs at the wavelength of 1440 nm, but, using smaller-sized crystals (3.4-nm diameter), the peak is at 980 nm. This size-dependent tunability can be used in hyperspectral visible image sensors. From a performance standpoint, short-wavelength infrared (SWIR) PbS CQD has achieved detectivities greater than $10^{12}$ Jones at room temperature, which is comparable with commercial InGaAs photodiodes [see Fig. 8.7(c)].

Among all the colloidal nanomaterials, mercury telluride (HgTe) CQDs have demonstrated wide spectral tunability from SWIR [10–13] up to THz

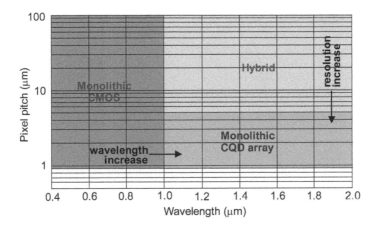

FIGURE 8.6 Positioning of a monolithic CQD array: higher wavelength than monolithic Si and greater resolution than hybrid alternatives (after Ref. [17]).

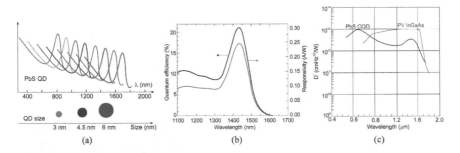

FIGURE 8.7 PbS CQD photodiode: (a) spectral tunability in dependence on dot size (after Ref. [17]), (b) quantum efficiency and responsivity, and (c) detectivity comparison of a PbS CQD photodiode with a commercial InGaAs photodiode (after Ref. [14]).

region [21]. For example, Fig. 8.8 shows the typical infrared absorption of different-sized thin-film HgTe dots in SWIR and MWIR regions, with cutoff wavelengths at 2.5 μm and 4 μm, respectively.

Figure 8.9 presents the construction of a HgTe CQD photovoltaic detector on a composite ITO/sapphire substrate. HgTe CQDs, with a final thickness of ~400 nm, are deposited layer-by-layer by drop-casting, and each layer is rinsed with a solution of 1,2-ethanedithiol, hydrochloric acid, and isopropyl alcohol. A solution of ~10-nm $Ag_2Te$ nanoparticles is then spun over the HgTe layer. To create a strongly p-doped HgTe CQD top layer, the $Ag_2Te$ nanoparticles are treated with the $HgCl_2$ solution, which liberates

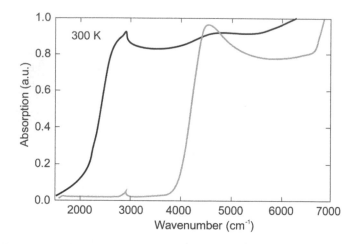

FIGURE 8.8 Infrared absorption of HgTe CQDs with cutoff wavelength at 2.5 μm (blue curve) and 4 μm (red curve) (after Ref. [22]).

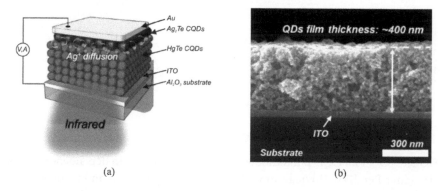

FIGURE 8.9 HgTe CQDs photovoltaic detector: (a) schematic detector diagram, (b) cross-sectional scanning electron microscopy image of the detector (after Ref. [22]).

the $Ag^+$ ions and immobilizes them as $AgCl_2$. All device fabrication is performed under ambient conditions.

Figure 8.10 shows steady improvement in performance over the past decade for MWIR HgTe CQD photodiodes. $D^*$ between $10^{10}$ to $10^9$ Jones at 5-μm was demonstrated for HgTe CQD devices, while maintaining a fast response time at thermoelectric cooling temperatures. The peak responsivity of 0.56 A/W is observed at about 160 K [22]. It is unlikely that CQD IR detectors will ever achieve the performance of the currently popular InGaAs, HgCdTe, InSb, and T2SL photodiodes. However, recent

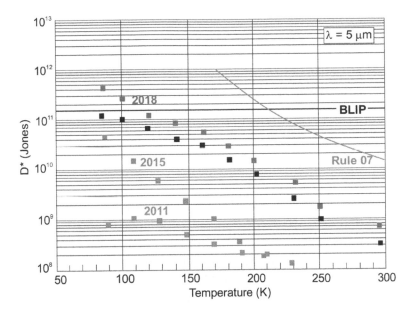

FIGURE 8.10 Progress in performance improvement during the past decade for HgTe CQD photodetectors at 5 μm. The filled red and black symbols are for detectors with and without optical enhancement of the absorption, respectively (after Ref. [10]).

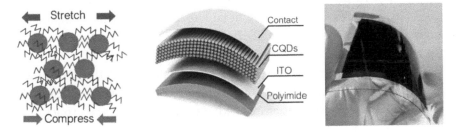

FIGURE 8.11 Flexible CQD photovoltaic detector: (a) flexible nanocrystal, (b) the device architecture of a flexible detector with a semitransparent top contact, and (c) the CQD layer deposited on flexible and unconventional substrates (after Ref. [22]).

demonstrations of low-cost SWIR and MWIR CQD-imaging arrays have heightened the interest in these devices.

Another unique advantage of CQD-based photodetectors noted in Table 8.1 is the mechanical flexibility. CQDs films can be stretched or compressed due to the fact that they are formed by closely packed nanocrystals linked by ligands (Fig. 8.11). The design and fabrication of flexible photodetectors have been demonstrated in many papers [e.g., 2,9,23].

Nowadays, imaging devices usually rely on planar focal plane arrays (FPAs). As has been described in Chapter 1, evolution of fourth-generation focal plane arrays is inspired by biological eyes. The human eye, despite a simple single-lens imaging system, is able to provide high-resolution imaging with adjustable zoom capability. The strategy of photodetector design, on a flexible substrate, can help in bending or folding, to shape the planar array into a hemispherical configuration.

## 8.3 IMAGING ARRAYS

It is expected that the successful implementation of a new class of CQD IR technology may match the broad impact of cheap CMOS cameras that are widely used today [24]. SWIR cameras, built on CQD thin-film photodiodes, and fabricated monolithically on silicon ROICs, have been launched [25,26]. Figure 8.12 shows two PbS CQD cameras fabricated by IMEC and SWIR Vision Systems. The IMEC's prototype imager has a resolution of 758 × 512 pixels and a 5-μm pixel pitch. The CQD photodiodes on the silicon substrate achieve an external quantum efficiency greater than 60% at 940 nm wavelength, and above 20% at 1450 nm, allowing uncooled operation, with a dark current comparable to that in commercial InGaAs photodetectors. The Acuros cameras, with a format up to 1920 × 1080 (2.1 megapixels, 15-μm pixel pitch), deliver 0.4 to 1.7 μm broadband, visible-to-SWIR, high-resolution images, at far superior cost points compared to InGaAs SWIR cameras (Fig. 8.12). In Table 8.3, the performance specifications of the Acuros camera are collated.

Figure 8.13 compares images taken with a visible and SWIR CQD camera during heavy rain over an ocean coast. Higher resolution images taken by the SWIR CQD camera were associated with reduced raindrop scattering [25].

(a)                                        (b)

FIGURE 8.12   New PbS CQD sensors integrated in camera modules with standard or SWIR lenses fabricated by (a) IMEC and (b) SWIR Vision Systems.

TABLE 8.3   Performance Comparison of SWIR Cameras

|  | Acuros™ CQD™ | InGaAs |
|---|---|---|
| Maturity | New | High |
| Spectrum | 400–1700 nm | 900–1700 nm |
| Quantum efficiency | 15% average | 65–80% |
| Resolution | 640×512 | 640×512 |
|  | 1280×1024 | 1280×1024 |
|  | 1920×1080 |  |
| Pixel pitch | 15 μm → ~2 μm | 15 μm → ~12 μm |
| Noise-equivalent irradiance | $6\times10^9$ photons/cm$^2$/s | ~$10^9$ photons/cm$^2$s |
| Operability | >99% | >99% |
| Cost | Low materials $ | Low materials $ |
|  | Low processing $ | Low processing $ |

FIGURE 8.13   Images obtained from the CMOS visible camera (left) and SWIR Acuros camera (right), during heavy rain over a Pacific Coast inlet, obscuring the 3-km distant shoreline. SWIR image enhances long-range visibility due to reduced raindrop scattering (after Ref. [25]).

At present, CQD cameras are used in newer applications that require high-definition, low-cost imaging on smaller pixels without extreme sensitivity [24–26]. It can be predicted that increasing the dot size, while maintaining a good mono-dispersion, carrier transport, and quantum efficiency, will improve/maintain low noise levels. Due to continuous development of deposition and synthesis techniques, much higher performances will be achieved in the future.

## 8.4   PRESENT STATUS OF CQD PHOTODIODES

In this section, we compare the performance of ideal P-i-N HgCdTe photodiodes with the present status of CQD photodiodes operated at room temperature. The theoretical estimation of HgCdTe photodiode performance, together with the experimental data, are described in detail in Section 3.3.

Figure 3.12(b) compares the temperature dependence of detectivity for different material systems with a cutoff wavelength of ~5 μm, namely commercially available HgCdTe photodiodes and HgTe CQD photodiodes. The experimental data collected are also included. The estimated detectivity values for CQD photodiodes are located below those for HgCdTe photodiodes. As is shown, at temperatures above 200 K, the theoretically predicted detectivity for HgCdTe photodiodes is limited by background. The semiempirical rule Rule 07, widely popular in the IR community as a reference for other technologies, is found not to fulfill primary expectations. Rule 07 coincides well with a theoretically predicted curve for an Auger-suppressed p-on-n HgCdTe photodiode, with a doping concentration in the active region equal to $10^{15}$ cm$^{-3}$. As indicated in Section 3.3, at the present stage of HgCdTe technology, the doping concentration is almost two orders of magnitude lower (mid-$10^{13}$ cm$^{-3}$).

All experimental data listed in Figs. 3.16 and 8.14 indicate a sub-BLIP photodetector performance by CQD photodetectors. Both figures also clearly show that the detectivity values of CQD photodetectors are inferior in comparison with those of HgCdTe photodiodes. Moreover, the theoretical predictions indicate a possible further performance improvement of HgCdTe devices after decreasing the i-doping level in P-i-N photodiodes. For doping levels of $5 \times 10^{13}$ cm$^{-3}$, the photodiode performance can be limited by background radiation in a spectral band above 3 μm. It is shown that, in this spectral region, the $D^*$ is limited, not by the detector itself, but by background photon noise at a level above $10^{10}$ Jones in the LWIR range (more than one order of magnitude above Rule 07).

In the past decade, considerable progress has been made in the fabrication of SWIR and MWIR CQD photodetectors, together with their integration into thermal imaging cameras. In spite of this, the performance of CQD photodetectors is inferior in comparison with HgCdTe photodiodes. It seems that the PbS CQD photodetectors, characterized by multispectral sensitivity and detectivity comparable with InGaAs devices (which are currently the most common in commercial applications), have been located at the best position in the IR-material family at the present time.

Between different material systems used in the fabrication of HOT LWIR photodetectors, only HgCdTe ternary alloy can fulfill the required expectations: low doping concentration ($10^{13}$ cm$^{-3}$) and high SRH carrier

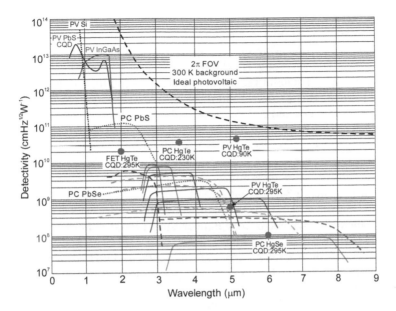

FIGURE 8.14  Room-temperature spectral detectivity curves of the commercially available photodetectors [PV Si and InGaAs, PC PbS and PbSe, HgCdTe photodiodes (solid lines, Ref. [27]). The experimental data for different types of CQD photodetectors are marked by dot points [3,9,28,29]. Spectral detectivity of new and emerging T2SL IB QCIPs (dashed lines) are also included [30]. PC: photoconductor; PV: photodiode.

lifetime (above 1 ms). In this context, it will be rather difficult to rival HgCdTe photodiodes with CQD photodetectors. The above estimates do, however, provide further encouragement for achieving low-cost and high-performance MWIR and LWIR HgCdTe focal plane arrays operated under HOT conditions.

## REFERENCES

1. P. Martyniuk, S. Krishna, and A. Rogalski, "Assessment of quantum dot infrared photodetectors for high temperature operation", *Journal of Applied Physics* **104**(3), 034314-1–6 (2008).
2. G. Konstantatos, "Colloidal quantum dot photodetectors", in *Colloidal Quantum Dot Optoelectronics and Photovoltaics*, pp. 173–198, eds. G. Konstantatos and E.H. Sargent, Cambridge University Press, Cambridge, 2013.
3. E. Lhuillier and P. Guyot-Sionnest, "Recent progress in mid infrared nanocrystal optoelectronics", *IEEE Journal of Selected Topics in Quantum Electronics* **23**(5), 6000208 (2017).

4. J.S. Steckel, J. Ho, C. Hamilton, J. Xi, C. Breen, W. Liu, P. Allen, and S. Coe-Sullivan, "Quantum dots: The ultimate down-conversion material for LCD displays", *Journal of the Society for Information Display* **23**, 294–305 (2015).

5. N. Ilyas, D. Li, Y. Song, H. Zhong, Y. Jiang, and W. Li, "Low-dimensional materials and state-of-the-art architectures for infrared photodetection", *Sensors* **18**, 4163 (2018).

6. G. Konstantatos and E.H. Sargent, "Solution-processed quantum dot photodetectors", *Proceedings of IEEE* **97**(10), 1666–1683 (2009).

7. N.C. Greenham, X. Peng, and A.P. Alivisatos, "Charge separation and transport in conjugated polymer/cadmium selenide nanocrystal composites studied by photoluminescence quenching and photoconductivity", *Synthetic Metals* **84**, 545–546 (1997).

8. D.S. Ginger and N.C. Greenham, "Photoinduced electron transfer from conjugated polymers to CdSe nanocrystals", *Physical Review B* **59**, 10622–10629 (1999).

9. F.P. Garcia de Arquer, A. Armin, P. Meredith, and E.H. Sargent, "Solution-processed semiconductors for next-generation photodetectors", *Nature Reviews Materials* **2**, 16100 (2017).

10. P. Guyot-Sionnest, M.M. Ackerman, and X. Tang, "Colloidal quantum dots for infrared detection beyond silicon", *Journal of Chemical Physics* **151**, 060901-1–8 (2019).

11. P. Guyot-Sionnest and J.A. Roberts, "Background limited mid-infrared photodetection with photovoltaic HgTe colloidal quantum dots", *Applied Physics Letters* **107**, 091115 (2015).

12. C. Buurma, R.E. Pimpinellaa, A.J. Ciani, J.S. Feldman, C.H. Grein, and P. Guyot-Sionnest, "MWIR imaging with low cost colloidal quantum dot films", *Proceedings of SPIE* **9933**, 993303 (2016).

13. C. Buurma, A.J. Ciani, R.E. Pimpinella, J.S. Feldman, C.H. Grein, and P. Guyot-Sionnes, "Advances in HgTe colloidal quantum dots for infrared detectors", *Journal of Electronic Materials* **46**(11) 6685–6688 (2017).

14. E.J.D. Klem, C. Gregory, D. Temple, and J. Lewis, "PbS colloidal quantum dot photodiodes for low-cost SWIR sensing", *Proceedings of SPIE* **9451**, 945104-1–5 (2015).

15. A. De Iacovo, C. Venettacci, L. Colace, L. Scopa, and S. Foglia, "PbS colloidal quantum dot photodetectors operating in the near infrared", *Scientific Reports* **6**, 37913 (2016).

16. M. Thambidurai, Y. Jjang, A. Shapiro, G. Yuan, H. Xiaonan, Y. Xuechao, G.J. Wang, E. Lifshitz, H.V. Demir, and C. Dang, "High performance infrared photodetectors up to 2.8 μm wavelength based on lead selenide colloidal quantum dots", *Optical Materials Express* **7**(7), 2336 (2017).

17. P.E. Malinowski, E. Georgitzikis, J. Maes, I. Vamvaka, F. Frazzica, J. Van Olmen, P. De Moor, P. Heremans, Z. Hens, and D. Cheyns, "Thin-film quantum dot photodiode for monolithic infrared image sensors", *Sensors* **17**, 2867 (2017).

18. A.D. Stiff-Roberts, "Quantum-dot infrared photodetectors: a review", *Journal of Nanophotonics* **3**, 031607 (2009).

19. M. Takase, Y. Miyake, T. Yamada, T. Tamaki, M. Murakami, and Y. Inoue, "First demonstration of 0.9 μm piel global shutter operation by novel charge control in organic photoconductive film", *Proceedings of the 2015 IEEE International Electron Devices Meeting* (IEDM), Washington, DC, USA, 7–9 December 2015.

20. "IMEC develops infrared thin-film sensor with 'record' pixel density", 26 Oct. 2019, https://optics.org/news/10/10/38

21. N. Goubet, A. Jagtap, C. Livache, B. Martinez, H. Portalès, X.Z. Xu, R.P.S.M. Lobo, B. Dubertrct, and E. Lhuillier, "Terahertz HgTe nanocrystals: Beyond confinement", *Journal of the American Chemical Society* **140**(15), 5033–5036 (2018).

22. X. Tang, M.M. Ackerman, and P. Guyot-Sionnest, "Colloidal quantum dots based infrared electronic eyes for multispectral imaging", *Proceedings of SPIE* **11088**, 1108803-1–7 (2019).

23. X. Tang, M.M. Ackerman, G. Shen, and P. Guyot-Sionnes, "Towards infrared electronic eyes: flexible colloidal quantum dot photovoltaic detectors enhanced by resonant cavity", *Small* **15**, 1804920 (2019).

24. S.B. Hafiz, M. Scimeca, A. Sahu, and D.K. Ko, "Colloidal quantum dots for thermal infrared sensing and imaging", *Nano Convergence* **6**, 7 (2019), https://doi.org/10.1186/s40580-019-0178-1

25. "SWIR Vision Systems", November 2018, https://ibv.vdma.org/documents/256550/27019077/2018-11-07_Stage1_1030_SWIR+Vision+Systems.pdf/

26. M. Chen, H. Lu, N.M. Abdelazim, Y. Zhu, Z. Wang, W. Ren, S.V. Kershaw, A.L. Rogach, and N. Zhao, "Mercury telluride quantum dot based phototransistor enabling high-sensitivity room-temperature photodetection at 2000 nm", *ACS Nano* **11**, 5614–5622 (2017).

27. *VIGO System S.A. Catalogue*, February 2018, https://vigo.com.pl/wp-content/uploads/2017/06/VIGO-Catalogue.pdf

28. C. Livache, B. Martinez, N. Goubet, J. Ramade, and E. Lhuillier, "Road map for nanocrystal based infrared photodetectors", *Frontiers in Chemistry* **6**, Article 575 (2018).

29. G. Konstantatos, "Current status and technological prospect of photodetectors based on two-dimensional materials", *Nature Communications* **9**, 5266 (2018).

30. A. Rogalski, P. Martyniuk, and M. Kopytko, "Type-II superlattice photodetectors versus HgCdTe photodiodes", *Progress in Quantum Electronics*. **68**, 100228 (2019).

# Final Remarks

Future civilian and military infrared and terahertz detector applications face many critical challenges. The main efforts are directed to decreasing the size, weight, and power consumption (SWaP) of infrared imaging systems by increasing the operating temperature of detector arrays. Nowadays, HgCdTe is the most widely used variable bandgap semiconductor for IR photodetectors, including uncooled operation, and stands as a for alternative technologies. However, in spite of sixty years of development history of the HgCdTe ternary alloy system, its ultimate HOT performance limit has not been achieved.

In Kinch's monograph [1] it is clearly shown that "the ultimate cost reduction for an IR system will only be achieved by the room-temperature-operation of depletion-current-limited arrays with pixel densities that are fully consistent with background- and diffraction-limited performance due to the system optics. This mandates the use of an IR material with a long S-R lifetime. Currently, the only material that meets this requirement is HgCdTe." Kinch predicted that large-area ultra-small pixel diffraction-limited and background-limited photon-detecting MW and LW HgCdTe FPAs, operating at room temperature, would be available within the next five years [2].

At present, the Rule 07 metric (specified in 2007 [3]) is not a proper approach for prediction of the HgCdTe detector and system performance, nor as a reference benchmark for alternative technologies (often used especially for type-II superlattices and colloidal quantum dots). In Section 3.3, it was shown, that, for sufficiently long SRH carrier lifetime in HgCdTe ternary alloy, what is experimentally supported by Teledyne at a doping level below $5 \times 10^{13} cm^{-3}$, the internal P-i-N HgCdTe photodiode current is suppressed and the detector performance is limited by the background radiation. HgCdTe photodiodes, operating in the longer-wavelength infrared range (above 3-$\mu$m), guarantee achieving more than one order of magnitude higher detectivity (above $10^{10}$ Jones) in comparison with the value predicted by Rule 07. This new benchmark [4], Law 19, corresponds exactly to the detector background-limited performance curve for room temperature. In this context, our paper evaluates a new class of emerging 2D material technologies for HOT infrared photodetectors.

At present, the performance of single room-temperature 2D material photodetectors, operating in infrared and terahertz spectral ranges, is comparable to that of standard commercial photodetectors (InGaAs and HgCdTe). Black phosphorus-arsenic alloys and noble TMD photodetectors have emerged as candidates for detection of long-wavelength infrared radiation with higher detectivities in comparison with commercial HgCdTe photodiodes. However, the instability of the black phosphorus surface, due to chemical degradation under ambient conditions, remains a major impediment to its prospective applications. More promising are stable noble TMD photodetectors like $PdSe_2/MoS_2$ heterojunctions, with record detectivity in the LWIR range at room temperature. However, their practical application lies in perfect material synthesis and processing. Due to mature HgCdTe technology, it is rather difficult for 2D material to compete with HgCdTe photodiodes. To achieve it, 2D material photodetectors have a long way to go.

The future applications imaging infrared systems require:

- Higher pixel sensitivity,

- Further increases in pixel count to above $10^8$ pixels (mosaicking may be used) with pixel size decreasing to about 5-μm for both cooled and uncooled LWIR applications,

- Cost reduction in imaging array systems through the use of the integration of detectors and signal processing functions (with much more on-chip signal processing) and less cooling sensor technology,

- Improvement in the functionality of imaging arrays through the development of multispectral sensors.

Small-pitch IR FPAs will require the development of larger effective ROIC well capacities per unit area, possibly faster optics than $f/1$, and improved hybridization technologies compared with those dominating currently in IR arrays fabrication. Leveraging deeply scaled CMOS process technology enables designers to miniaturize pixel pitch and/or increase on-chip processing capability, depending on application-specific needs. Array sizes will continue to increase but perhaps at a rate that falls below the Moore's Law curve. An increase in array size is already technically feasible. However, the market forces that have demanded larger arrays are not as strong now that the megapixel barrier has been broken.

The first 2D material-based imaging sensors operating in the visible and short wavelength infrared region have been demonstrated [5]. Dual band vertical GaSe/GaSb heterostructure linear array (16×1) was fabricated by the MBE growth technique in 2017 [6]. Also 388×288 20-µm pixel graphene-QD-based CMOS-integrated sensor has since been demonstrated [7]. Graphene is particularly attractive due to its compatibility with silicon and its monolithic integration with CMOS-integrated circuits. However, development of photodetectors toward high performance, high integration, and a large focal plane is particular critical for practical applications of 2D photodetectors in industry. In a way, graphene has been a victim of its own success, getting people overly excited about it and attracting unrealistic expectations. High-quality 2D materials are the promising next-generation alternative, although it will take several decades of research, development, and, most importantly, billions of dollars of investment at the national and international level to become the ultimate benchmark for standard electronic materials and devices. From the other side, the evolutionary path of Si technology, driven by Moore's Law of Scaling, seems to be narrowing and fast approaching an end, simply due to the fundamental limitations of Si at the atomic scale [8]. In fact, it is unlikely that 2D technology will supplant Si, but instead it may coexist with Si technology [9]. This, too, will require significant research and resource investment in large area growth of 2D materials at temperatures compatible with silicon-based technology.

Next, we follow Ref. [10] to compare different detector technologies existing on the global market, with emerging 2D photodetectors. Table 1 provides a snapshot of the current state of development of LWIR detectors fabricated from different material systems. Note that TRL means technology readiness level. The highest potential TRL (ideal maturity) achieves a value of 10. The highest actual level of maturity (TRL = 9) is credited to HgCdTe photodiodes and microbolometers, with QWIPs offering a slightly lower maturity, with TRL = 8. The type-II $A^{III}B^V$ superlattice structures have great potential for LWIR spectral range applications, with performance comparable to HgCdTe for the same cutoff wavelength. Considerable progress toward mature superlattice and barrier detector technologies, including their commercialization, has been observed in the

TABLE 1 Comparison of LWIR Existing State-of-the-Art Device Systems for LWIR Detectors

| **Maturity level** | Bolometer | HgCdTe | QWIP | Type-II SLs | 2D Materials |
|---|---|---|---|---|---|
| | TRL 9 | TRL 9 | TRL 8 | TRL 7 | TRL 1–2 |
| Status | Material of choice for applications requiring medium to low performance | Material of choice for applications requiring high performance | Commercial | Research and development | Research |
| Operating temp. | Un-cooled | Cooled | Cooled | Cooled | Un-cooled |
| Manufacturability | Excellent | Poor | Excellent | Very good | Very poor |
| Cost | Low | High | Medium | Medium | ? |
| Prospect for large format | Excellent | Very good | Excellent | Excellent | ? |
| Availability of large substrate | Excellent | Poor | Excellent | Very good | Very good |
| Military system examples | Weapon sight, night-vision goggles, missile seekers, small UAV sensors, unattended ground sensors | Missile intercept, tactical ground and air-borne imaging, hyper spectral, missile seeker, missile tracking, space-based sensing | Being evaluated for some military applications and astronomy sensing | Being developed in university and evaluated in industry research environment | Military wearable devices |
| Limitations | Low sensitivity and long- time constants | Performance susceptible to manufacturing variations. Difficult to extend to >14-μm cutoff | Narrow bandwidth and low sensitivity | Requires a significant investment and fundamental material breakthrough to achieve maturity | Very limited yields, poor reproducibility and scalability; requires radical breakthrough to achieve mature technology |

*(Continued)*

TABLE 1 (CONTINUED)  Comparison of LWIR Existing State-of-the-Art Device Systems for LWIR Detectors

| | Bolometer | HgCdTe | QWIP | Type-II SLs | 2D Materials |
|---|---|---|---|---|---|
| **Maturity level** | TRL 9 | TRL 9 | TRL 9 | TRL 7 | TRL 1–2 |
| Advantages | Low cost and requires no active cooling, leverages standard Si manufacturing equipment | Near theoretical performance, will remain material of choice for minimum of the next several years | Low-cost applications. Leverages commercial manufacturing processes. Very uniform material | Theoretically better than HgCdTe, leverages commercial III-V fabrication techniques | Performance of single devices comparable with standard devices |

*Note:* TRL – technology readiness level.

past decade, with 2D material detector technology being at a very early stage of development (TRL = 1–2).

Development of a scalable 2D material detector arrays concept, with appropriate fabrication tools, can be a noteworthy indicator that the 2D materials can enable the development of new detector systems. The combination of scalability, the prospects for integration with Si-platforms, and the potential for implementing flexible devices can make 2D material attractive for future generations of detection systems. However, at the present stage of detector technology, it will be rather difficult to improve their position in a mature global market detector family in the near future. From an economical point of view and a future technology perspective, an important aspect concerns industry fabrication of detector arrays with high operability, spatial uniformity, temporal stability, scalability, producibility, and affordability. All these aspects are at a stage of infancy in terms of manufacturability.

## REFERENCES

1. M.A. Kinch, *State-of-the-Art Infrared Detector Technology*, SPIE Press, Bellingham, 2014.
2. M.A. Kinch, "An infrared journey", *Proceedings of SPIE* **9451**, 94512B (2015).
3. W.E. Tennant, D. Lee, M. Zandian, E. Piquette, and M. Carmody, "MBE HgCdTe technology: A very general solution to IR detection, described by 'Rule 07', a very convenient heuristic", *Journal of Electronic Materials* **37**, 1406–1410 (2008).
4. D. Lee, P. Dreiske, J. Ellsworth, R. Cottier, A. Chen, S. Tallarico, A. Yulius, M. Carmody, E. Piquette, M. Zandian, and S. Douglas, "Law 19 – The ultimate photodiode performance metric", Extended Abstracts. The 2019 U.S. Workshop on the Physics and Chemistry of II-VI Materials, pp. 13–15, 2019.
5. M. Long, P. Wang, H. Fang, and W. Hu, "Progress, challenges, and opportunities for 2D material based photodetectors", *Advanced Functional Materials* **29**, 1803807 (2018).
6. P. Wang, S. Liu, W. Luo, H. Fang, F. Gong, N. Guo, Z.-G. Chen, J. Zou, Y. Huang, X. Zhou, J. Wang, X. Chen, W. Lu, F. Xiu, and W. Hu, "Arrayed van der Waals broadband detectors for dual-band detection", *Advanced Materials* **29**, 1604439 (2017).
7. S. Goossens, G. Navickaite, C. Monasterio, S. Gupta, J.J. Piqueras, R. Pérez, G. Burwell, I. Nikitskiy, T. Lasanta, T. Galán, E. Puma, A. Centeno, A. Pesquera, A. Zurutuza, G. Konstantatos, and F. Koppens, "Broadband image sensor array based on graphene–CMOS integration", *Nature Photonics* **11**, 366–371 (2017).

8. *International Roadmap for Devices and Systems™. 2018 Update. More Moore.* https://irds.ieee.org/images/files/pdf/2018/ 2018IRDS_MM. pdf

9. N. Briggs, S. Subramanian, Z. Lin, X. Li, X. Zhang, K. Zhang, K. Xia, D. Geohegan, R. Wallace, L.-Q. Chen1, M. Terrones, A. Ebrahimi, S. Das, J. Redwing, C. Hinkle, K. Momeni, A. van Duin, V. Crespi, S. Kar, and J.A. Robinson, "A roadmap for electronic grade 2D materials", *2D Materials* **6**, 022001 (2019).

10. *Seeing Photons: Progress and Limits of Visible and Infrared Sensor Arrays,* Committee on Developments in Detector Technologies; National Research Council, 2010, http://www.nap.edu/catalog/12896.html

# Index